A Study of the Fukushima Daiichi Nuclear Accident Process

Michio Ishikawa

A Study of the Fukushima Daiichi Nuclear Accident Process

What caused the core melt
and hydrogen explosion?

 Springer

Michio Ishikawa
Former Professor of Hokkaido University
Tokyo, Japan

Kosho Fukushima Genshiryoku Jiko - Roshin Youyu Suiso Bakuhatsu ha Dou Okottaka-
© Michio Ishikawa 2014
Published by The Denki Shimbun, Japan Electric Association Newspaper Division
1-7-1 Yurakucho, Chiyoda-ku, Tokyo, Japan

ISBN 978-4-431-55542-1   ISBN 978-4-431-55543-8   (eBook)
DOI 10.1007/978-4-431-55543-8

Library of Congress Control Number: 2015947564

Springer Tokyo Heidelberg New York Dordrecht London
© Springer Japan 2015

Printed on acid-free paper

Springer Japan KK is part of Springer Science+Business Media (www.springer.com)

Foreword

Dr. Michio Ishikawa, the author of this book, is a leading authority in a field of nuclear safety engineering in Japan. He made great achievements contributing to the worldwide study of nuclear safety as a researcher at the former Japan Atomic Energy Research Institute, nurtured numerous young engineers as a professor of Hokkaido University, served as a government consultant in the licensing of various atomic power plants throughout Japan, participated as the Japanese representative in developing the international nuclear safety standards of the International Atomic Energy Agency (IAEA), and also served as the first president of the Japan Nuclear Technology Institute. In these positions, he has come to be known worldwide as one of the leading authorities on atomic energy.

As I recall, I started working with him at the time of the JCO (a uranium conversion facility) criticality accident in 1999, when I was serving as the Minister of Education as well as the Director General of the Science and Technology Agency. This was followed by working together again as the chairman and vice chairman of the Society for Promotion of Nuclear Safety and Utilization, which was founded in the summer of 2013. It was on that basis that I came to have a great interest in the fact that Dr. Ishikawa was writing a book that would unravel the mystery of the complex phase of the accident at the Fukushima Daiichi Nuclear Power Station. I was quite impressed reading his draft to see his complete clarification of all the accident phenomena that occurred in the nuclear reactors, which have been considered very difficult to understand until now. The book provides a tremendous amount of analyses and studies that should be made known to the world, and I felt an urge to seek the honor of authoring the recommendation of the book when it was published.

According to Dr. Ishikawa, in the particular loss of coolant accident that occurred in the light water reactors, the fuel rods retained their original shapes thanks to heat dissipation due to steam cooling and radiation under high temperatures, as well as to the tough oxidation skin formed on the surface of the cladding tube, and were broken down into pieces when they were quenched and dropped to the bottom of the reactor to cause a core disruption. The continuous nuclear residual heat (decay heat) emitted from the fuel alone was not sufficient to readily cause melting; rather, the fuel rods melt occurred only when the intense heat generated by an oxidation reaction between the cladding tube material (zirconium) and water became accelerative. However, said reaction forms an eggshell (crust) between the molten matter and water, which protects the molten core inside the eggshell, thus preventing various thermal violence, such as a steam explosion. Dr. Ishikawa explains such a chain of events based on the results of the fuel rod behavior test conducted as a part of the safety proof study conducted by Japan, the U.S., and Germany in the 1970s, in which he personally participated, as well as on the results of the survey on the conditions of the molten core left inside the nuclear reactor pressure vessel in the Three Mile Island (TMI) nuclear power station accident that occurred in 1979.

As a result of the analysis of the Fukushima accident based on these findings, Dr. Ishikawa clearly answers such questions as: why there were timing differences in the core melts among Units 1, 2, and 3; why a large amount of hydrogen generated in the core melts caused explosions outside the containment vessels and lead an extreme damage to the reactor buildings; and why the explosion occurred only on the top floor in Unit 1. Why there was no explosion in Unit 2; why the explosion center moved from the top floor to the lower floors in Unit 3, while it moved from the bottom floors toward the top floor in Unit 4. Dr. Ishikawa wrote. "In the TMI accident the core melt occurred within a short period. This is because a large amount of water was supplied by the primary coolant pump, thus causing a hydrogen explosion in the containment vessel. In the case of Fukushima Daiichi, however, there was no means of supplying water other than fire extinguishing pumps, thus delaying the water–metal reaction due to the shortage of water, although eventually a core melt occurred." Dr. Ishikawa explains this long chain of events unit-by-unit without leaving any contradiction, following every detail of the published measurement data at the time of the accident, without leaving any contradiction. These are matters that people around the world who are associated with nuclear power eager to know.

Dr. Ishikawa's knowledge of nuclear accidents is truly outstanding. As for the Chernobyl nuclear power plant accident, which ended up spraying a huge amount of radioactive materials throughout the entire world due to a lack of containment vessel, he reveals generally unknown fact that the graphite fire that occurred in the early stage of the accident in fact had the effect of cooling the core, that the fuel melt occurred due to the decaying heat only after the graphite moderator had burnt itself out and increased the radioactive release. The actual amount of radioactive release in the Fukushima accident, where there were containment vessels, was only one-seventh that of the Chernobyl accident, or one-fifteenth in terms of release per unit power output. In addition, according to his analysis, the core melt could have been prevented and the damaged core could have been cooled safely had the pressure increase been prevented successfully by venting the containment vessel, and had a sufficient amount of water been supplied simultaneously with the reactor pressure reduction. He concludes that a containment vessel vent of the boiling water reactor (BWR) is extremely effective when the radioactive gas is released after passing through the water of the wet well. He also concludes that it would not have even been necessary to evacuate the residents of the surrounding areas had the vent been carried out successfully. It is extremely regrettable that this was not known until the accident occurred, but it is also very encouraging for those of us who are involved in nuclear power engineering and should bring great relief to citizens who are concerned about the safety of nuclear power generation.

Dr. Ishikawa criticizes the rash view of the Accident Investigation Board of the Diet, which assumes that the piping of Unit 1 was broken by the earthquake even before the tsunami struck, as a big mistake. Moreover, he argues that the new control standard drawn up by the Nuclear Regulatory Commission, which was established after the Fukushima accident, neglects the basics of safety design standards having been formed as a result of the deep insights and careful evaluations over many years of nuclear safety engineers worldwide. It is based on a perception that the strictest rule in the world is the best choice, with no heed to overall optimization. For example, he astutely points out that an abnormally high seawall can be harmful once it is overcome by waves as it takes a long time to drain the water retained inside the seawall, and that it is not necessary to install a filter vent based on the analysis for the Fukushima accident. He is concerned that the demands made in the safety design fortification are imbalanced from the practical standpoint and that it may even reduce safety as a result.

As a proposal for the future, he claims that it is important to seek diversity in the countermeasures against natural phenomena as well as against terrorist attacks, in addition to safety designs that have hitherto been considered. Furthermore, he emphasizes that it is most essential to seek Mitigation Safety System against Disasters (MISSAD) such as the distributed installation of portable systems, including help from outside of the power station.

It is worth listening to his proposals that we must review quickly the 1 mSv/y rule which has become the de facto standard for the evacuees' return to their homes, that we should combine the establishment of a research center with the decommissioning work, and that we should accumulate worldwide knowledge to contribute to the recovery and rebirth of Fukushima. He argues, that the current status, such as hinting at the fact that there has been no death caused by the exposure to radiation is only met by harsh criticisms or that no evacuees are allowed to return to their homes, is itself the problem, that the current sense of hopelessness, e.g., the evacuees' deaths caused by the unnecessary hardships of evacuation and that the delay of evacuees' returns to their homes could have been avoided, had reasonable consideration been given to the IAEA's evacuation standard applied to the true state of exposure dose. These points should be taken into consideration in the nuclear disaster prevention measures to be adopted in the future.

The results of Dr. Ishikawa's analyses and studies prove that a reconstruction of the safety of nuclear power generation is feasible. It is good news for us to know that we can continue to use nuclear power safely when we face the reality that energy resource-poor Japan cannot be without nuclear power.

Now, the mode of core melts and explosions at Fukushima Daiichi Power Station are completely deciphered by this book. It takes a unique form of description, stepping back and forth in some cases as in a textbook so as to make it easier for general readers to understand. However, given that the Fukushima Daiichi accident was so complicated and the description of a nuclear disaster involves such expertise, even a nuclear engineer or researcher may have difficulty in understanding completely the content of this book. Therefore, I suggest that general readers should try to avoid being entrapped in every detail and just read it through. That alone should provide a reader a substantial step forward in understanding this nuclear reactor accident and disaster. However, students of nuclear power safety should read it through completely and understand its contents thoroughly. I believe that international professionals must have been waiting for this kind of contribution from Japan concerning the Fukushima accident. Japan should explain this painful experience to the world in order to contribute to the improvement of nuclear safety. I believe that this book will serve greatly towards that purpose, and I am sure that the study of the Fukushima accident by this book will surprise experts around the world.

Former President of The University of Tokyo   Akito Arima
Tokyo, Japan

Preface

Why I Decided to Write This Book

"Red-hot core fuel rods are melting down and turning into round balls of molten substance, dropping on the bottom of the pressure vessel. The heat caused the molten core and water in its vicinity to react to generate hydrogen, which in turn caused a hydrogen explosion to damage the reactor building. The red-hot molten core further penetrated through the bottom of the pressure vessel to drop on the floor of the containment vessel and melted the concrete floor of the containment vessel as well. Luckily, because the decay heat from the core subsided, the molten substance penetrating through the containment vessel seems to have been just barely prevented."

Such was the typical expression used by NHK (Nippon Hoso Kyokai, the public TV network in Japan) in their reports of what happened in the accident in Tokyo Power Electric Company's (TEPCO's) Fukushima Daiichi Nuclear Power Station. I suppose that the image most readers of this book have of the Fukushima accident is basically the same as this. It is not just the readers. The understanding that the majority of the people related to the nuclear accident have in their minds is probably not that different from the above. However, such an understanding is far from the truth.

As proof of this, after 3 years have passed since the accident, we have not seen any explanation presented by the government, nuclear power industry, or TEPCO as yet on how the core melt and the hydrogen explosion occurred, which is the crux of the accident. It is true that the phenomenon that occurred in the accident in question is very complex, but the reason that they failed to explain it logically is because there is an error in the image described above.

This book tries to correct the error and to analyze the specific phenomena that occurred in Fukushima Daiichi's Units 1 through 4. The data used here are strictly limited to indisputable accident phenomena and the actual measurement data that can be found in *The Fukushima Daiichi Nuclear Power Station Disaster* reported in June 2012 by TEPCO (original Japanese version; an English version was issued in March 2014). The rest consists of my studies of the factual data. My studies are based on the analyses of the accident phenomena of the Three Mile Island (TMI) Nuclear Reactor in the United States and the Chernobyl Nuclear Power Plant in the Ukraine (then U.S.S.R.) as well as the results of the safety experiments conducted jointly between the U.S., Germany, Japan, and France. In other words, this book is the result of deductions based on, and studies of, the actual facts.

The decision to write this book was triggered when I was asked to attend a meeting held in November 2012 with the U.S. Science Academy's team investigating the accident. The meeting's primary purpose was to investigate the facts of the accident, but the questions raised in the meeting were almost all about confirming the facts and the comparisons between them and the results of the analyses using well-known computational codes. It seemed to me that this was an exercise to check the computational codes, which is somewhat different from a discussion seeking to clarify what really happened in the accident.

Since the well-known computational codes were created based on the past experiences, it was obvious that they might not fit perfectly with what really happened in the unexperienced accidents. In particular, the Fukushima accident was extremely complex and contained various phenomena that had not been incorporated in the computational codes. Using such codes, if we tried to fine-tune the input too radically in order to make the calculation match the measurement data, the overall coherence of the accident would be lost. I guess what they meant to do in the meeting was to identify such discrepancies so that they could modify the computational codes for the future. Consequently, no clear-cut explanation has been provided for what really happened in the Fukushima accident.

The inability to provide a clear explanation is due to the use of a feeble investigative method that relies on computer calculations without having conducted a physiochemical clarification of the accident phenomena. I thought that this was a common problem of both the Japanese and American teams. I had not gotten involved in investigating the accident, thinking that there was no need for an old fellow like me to do so, but the meeting between the Japanese and American teams above changed my thinking. If the younger people we had nurtured did not have the ability to do so, I must take the helm, I thought. Studying again was not easy for an old man.

Although from the time of the accident I had a rough sense of how it progressed and how the various phenomena could be explained, checking each phenomenon required a lot of work. After about 6 months of staring at the reports and data trying to understand the various phenomena, I finally thought I generally understood the accident, so I started to write. However, during the course of writing, I often came up with questions concerning my interpretation of a fact, and the answers affected the overall understanding of the incident sometimes, so that I had to go through days of agony of further investigation and rewriting to eliminate contradictions. I now believe that I wrote everything I needed to write about, without any major mistakes.

This book consists of Part I, which is a study of the mechanisms of the core melts and hydrogen explosions that occurred in the Fukushima Daiichi nuclear power station accident, and Part II, which compiles my thoughts on the issues of the general safety of atomic power and the reconstruction of Fukushima, including the causes and effects of the accident. The main purpose of this book is condensed into Part I, so I hope that readers spend time to read it carefully.

The bases of my studies presented in Part I are the facts by the aforementioned TMI and Chernobyl accidents, as well as the results of a series of experiments called the fuel rod behavior test. It was a part of the joint safety research studies concerning light water reactors conducted for about 10 years since 1975 mainly by the U.S., Germany, and Japan. Their essential aspects are described in Chap. 1. Because my studies of the Fukushima accident are essentially based on these basic findings, readers are encouraged to read Chap. 1 carefully to grasp the ideas. Following my thoughts should not be that tough.

Chapter 2 deals with the performances (core melts and hydrogen explosions) of the nuclear reactors of Fukushima Daiichi's Units 1 through 3; it is the centerpiece of the book. It is meaningless for people to try to have an overall perspective of the accident without first examining the data and accident phenomena of Fukushima carefully one by one. Therefore, you may find some parts of the book annoying, as similar narrations will come up again and again. You may skip those repetitions if you wish. What is important is for you to grasp the outline of the accident; there is no need for you to be tied up in the details. A summary is provided at the end of each segment of Chaps. 1 and 2 in particular so that you can grasp the outline of each accident to prepare you for the next segment. The Fukushima accident is that complicated and difficult to understand.

However, those of you who are trying to be the young lions of the next generation in the field of nuclear power should read Chaps. 1 and 2 carefully while sweeping away your sleepiness. If you do so, you should be able to see automatically how wrong the core meltdown story I presented at the start of this preface is. And that knowledge will contribute to the improvement in the safety of nuclear power in the future.

Chapter 3 is my study of the explosion that occurred in Unit 4 of Fukushima Daiichi.

Part II describes my thoughts on various issues about the Fukushima accident from the technical point of view. It contains some outspoken comments which I believe will be useful for the future of the nuclear safety and the reconstruction of Fukushima. I hope that this book is useful for such purposes.

Now let's move on to Chap. 1 of Part I, the basis of understanding the accident phenomena.

Tokyo, Japan   Michio Ishikawa

Contents

Part I
What Caused the Core Melt
and Hydrogen Explosion?

Chapter 1
Three Mile Island Nuclear Power Station Accident

1.1 Shape of the Molten Core

There is something we have to learn before we get down to our main subject. That is the core melt accident that occurred in the Unit 2 reactor of the Three Mile Island (TMI) Nuclear Power Plant in the United States, a precedent for our case. Why? Because what happened there is very close to the core melt that occurred in the Fukushima Daiichi Nuclear Power Plant.

The TMI accident occurred 4 a.m., March 28, 1979, approximately 35 years ago. Almost all of the molten core has been removed from the reactor and transported to the Idaho Nuclear Laboratory (INL) for storage. And yet, it seems that a special permit is required to enter the TMI because it is still highly radioactively contaminated.

Since this is the United States, a scientific and technological powerhouse, it goes without saying that the molten core was studied with utmost thoroughness before it was removed from the reactor. It should be noted here that the investigation was conducted under the international cooperation in which Japan participated. Although the details of the study were shared with Japan as well, unfortunately It seems to me that not many people who studied it closely.

Younger Japanese engineers rested on the laurels of the safety record that their predecessors built. They were negligent in studying about the actual accident. This was evident from their immediate reactions at the Fukushima Daiichi accident site. If only they had studied better what happened in TMI and Chernobyl, their reactions at the site as well as what they communicated with the public would have been very much different.

TMI reactors are light water reactors which are also called pressurized water reactors (PWR). These reactors are, incidentally, primarily favored by power companies of Western Japan, typically Kansai Electric Power Co., Inc. Although they are different from the boiling water reactors (BWR) of Fukushima Daiichi in how they produce the high pressure steam that is used for driving turbine generators, when it comes to the structure of the reactors they are almost the same. In particular,

© Springer Japan 2015
M. Ishikawa, *A Study of the Fukushima Daiichi Nuclear Accident Process*,
DOI 10.1007/978-4-431-55543-8_1

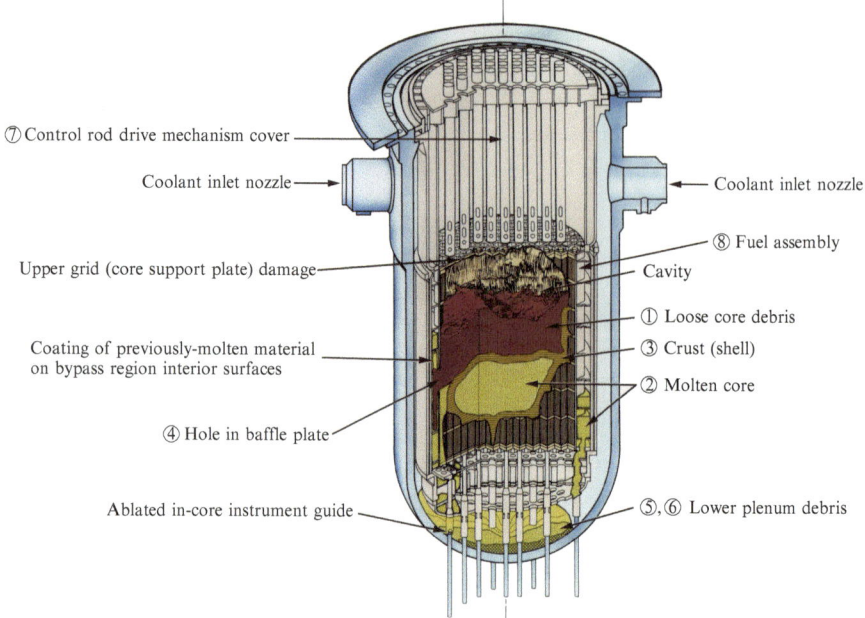

Fig. 1.1 TMI end-state configuration (Source: U.S. NRC site (http://www.nrc.gov/images/read-ing-rm/photo-gallery/20071114-006.jpg))

there is a lot of similarity between the PWR and the BWR in terms of core structures, materials, and physical dimensions, which are relevant to the accidents. In terms of the fuel rods, which played the key role, while the physical dimensions are slightly different, the materials and structures are almost identical.

Therefore, in order to understand the core melt condition of the Fukushima accident, understanding the TMI accident, which has been thoroughly studied is the quickest approach. So, let us look at closely what happened in the TMI's core melt.

First, let's look at Fig. 1.1. This is a cross-sectional view of the TMI pressure vessel focusing on the molten core.

The area marked as ① (the reddish black part) in the upper portion of the core represents fuel debris, which are the bits and pieces of the broken fuel rods remained at the top part of the core not completely molten, producing piles of wreckage. This kind of deposition of fuel is generally called "debris." As we progress in the chapter, you will see that the process of forming the debris plays a key role in clarifying the core melt.

The next area ② is the molten core. The thin crust (shell) ③ covering the molten core is an egg-shell-like object formed as the molten core was cooled by water, and it is sufficient for you to understand that it's an object with a physical property as hard as that of a casting product. During the accident, this crust (shell) served a role like that of the bottom of a pan, separating the squashy molten core, which kept generating heat, from the coolant that was on the outside of the molten core. The top surface of this crust (shell) ③ is flat because it was pressed down by the weight of the debris ①, which was lying on top of it.

The squashy molten core (shown on the left side of the drawing) moves sideway as it is pushed by the weight of the debris, and produced holes ④ in the baffle plate, which is made of a thin stainless steel plate to cover the core. It is said that a portion of the molten core was squeezed out of the baffle plate and formed a layer of solid substance ⑥ at the bottom of the pressure vessel.

According to another theory, the debris became solidified as it was cooled in the coolant path to form ball-like objects ⑥ with diameters of 10–15 cm that dropped and accumulated in the lower spherical plenum of the pressure vessel. In any event, it is true that a portion of the molten core was pushed out and accumulated at the bottom of the pressure vessel.

What is interesting as we look at this drawing is that the molten core seems to form an "egg-shell" instantaneously as it contacts with water. What is suggested here is the existence of a marvelous phenomenon: as soon as it was formed, this egg-shell served as an unexpectedly strong barrier between the molten core whose temperature must have been at least as high as 2,000 °C, and the surrounding water whose temperature was only several 100 °C. This is obvious not only from the existence of the molten substance in the middle of the core but also from the existence of the molten core balls at the bottom of the pressure vessel.

Had this egg-shell not been formed, the contacts between the high-temperature molten core and water could have caused violent thermal disturbances everywhere in the core and would have left evidences of the accident throughout the reactor. Yet, as I show you later, there is almost no trace of thermal disturbances due to the direct contact between the molten core and water inside the reactor. It seems that the core melt occurred quietly and then cooled and solidified quietly. The chemical nature of a molten core appears to have a very interesting and also very important nature that it produces an egg-shell ③ within a short period of time as it contacts with water.

The reason I wish to stress this characteristic here is that it is generally believed that a core melt occurs at a super high temperature, close to the melting point of uranium dioxide (UO_2), i.e., 2,880 °C, or higher, so that there are many people who believe that its coming into contact with water would cause a steam explosion. This is far from the real picture. Even among nuclear engineers, it is not difficult to find many who believe such a story; however, it is only a modern superstition.

If a steam explosion occurs in a nuclear reactor, it is only when, as in the Reactivity Initiated Accident, a high temperature substance (fuel) melts and evaporates due to a massive instantaneous generation of heat and causes an extremely large heat transfer surface. Moreover, it is limited only when the surrounding coolant (water) is at a lower temperature, e.g., the normal temperature. When the core temperature increases gradually to cause a core melt eventually as in the cases of TMI and Fukushima Daiichi, the steam explosion does not occur; and in fact they did not occur. The existence of the egg-shell is the proof of that.

Let me go back to where I vaguely described the core temperature as "at least as high as 2,000 °C," which I did for a reason. People generally believe that the core melt temperature is the melting point of uranium dioxide (UO_2), the fuel. However, when a fuel rod reaches a high temperature approximately 2,000 °C, contacts occur on the boundary between the uranium dioxide of the fuel and the zirconium (Zr) alloy (zircaloy[1]) of the cladding tube, and form a mixed molten substance consisting uranium, zirconium and oxygen.

This reaction is an extremely complex and highly technical. However, it is known that this mixed molten substance melts at a much lower temperature than the melting point of uranium oxide. The melting temperature of the core varies with the constituent elements and their constituting ratios, but it is said that it was 2,200 °C or thereabout in the case of TMI. In other words, TMI's core melt occurred at a temperature of approximately 2,200 °C. Let us move on making the speculation that the mixed melting temperature of the core in case of Fukushima Daiichi was not too different from the above.

Since we are on that subject, let me say a few words about Fig. 1.1 in order to help your understanding. The rod-like objects ⑦ appearing in the space above the TMI's core are the covers of the drive mechanism that propels the control rods up and down. This figure shows that these multiple covers survived, retaining their original shapes intact despite the accident. This is the evidence that the thermal disturbance of the core melt did not reach to this point and also that no destructive force such as an explosion occurred. The point is that, as I have written above, there is no evidence of a major thermal disturbance anywhere in the reactor in case of the TMI accident. I suppose that this may contradict with common sense, but it seems that a core melt occurs in a relatively localized way.

In case of the TMI accident, the control rods were all inserted in the core. The material used in the control rods were low-melting point silver-indium-cadmium alloy (melting point: approximately 800 °C) and the melting point of stainless steel that covers the control rods is also 1,450 °C. It seems that the control rod material melted relatively earlier and gathered around the lower half of the thin crust (shell) ③ that surrounded the molten core. This fact clearly proves that the core would never have reached recriticality even though the control rods were unevenly distributed within the molten core.

Next, a slender rod-like object ⑧ is shown on the right side of the molten core. This is an illustration of a fuel assembly that did not melt.

The heat generated in a reactor is higher in the center of the core and lower in the periphery. This is due to the fact that neutrons leak outside in the peripheral area of a reactor, so that the fission reaction rate is lower in the periphery. The fuel assemblies in the edge sections that survived without melting prove the particular thermal distribution. Similarly, as the bottom part of the core is also on the periphery, the fuel rods in the bottom part do not melt as their temperatures are lower, and are supporting the "egg-shell," in the form of randomly placed snaggle-teeth.

---

[1] Zircaloy is a kind of zirconium alloy. As it absorbs thermal neutrons, which causes fission, only to a limited degree, zirconium is used for fuel cladding tubes, channel boxes, etc. I refer to zirconium and zircaloy later in this book: I use the former to focus on the characteristics of zirconium atoms, and the latter to focus on the characteristics of the alloy used for fuel cladding tubes, etc.

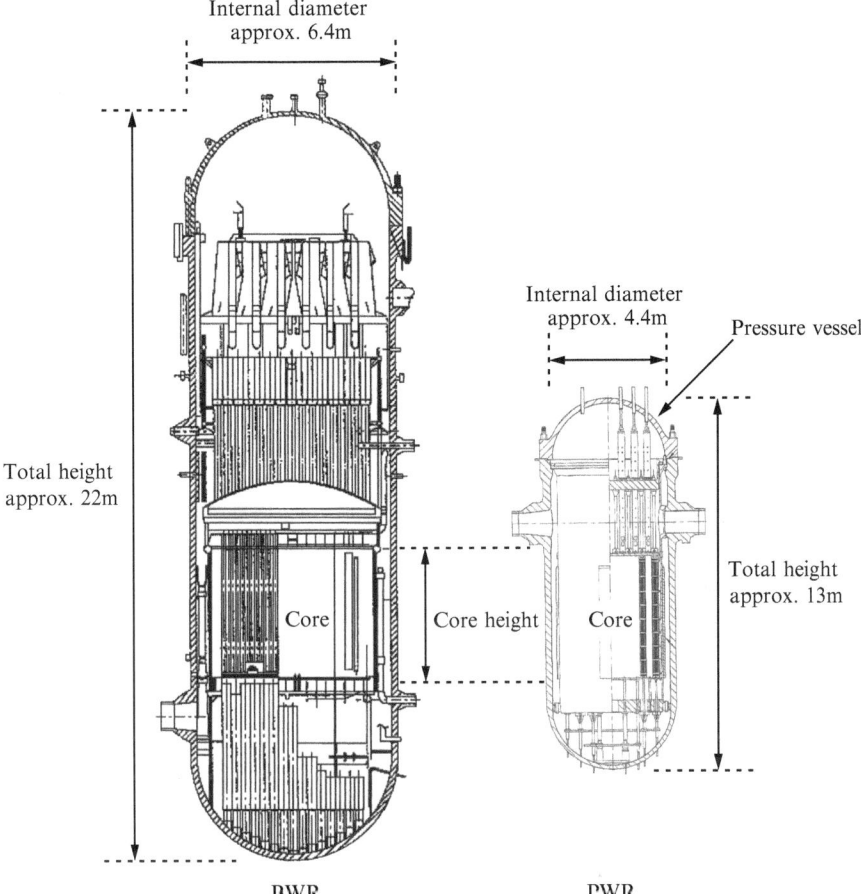

Fig. 1.2 Internal structures of BWR and PWR pressure vessels (Source: BMR: from TEPCO "Fukushima Nuclear Accident Investigation Report" PWR: from "Application documents for Tomari-3 reactor installment license")

This concludes the description of the molten core illustration shown in Fig. 1.1.

Let me briefly describe the difference between the core structure of the Boiling Water Reactor (BWR), which is the reactor type of the Fukushima accident, and that of the Pressurized Water Reactor (PWR). Please look at Fig. 1.2. As the lengths of the fuel rods for PWRs and BWRs are both approximately 4 m, and thus the height of the core is identical in both cases.

Let us compare the two types of pressure vessels having the cores of the same height. The BWR pressure vessel is generally larger in diameter and longer in length than the PWR pressure vessel. The reason that the BWR pressure vessel has a longer bottom section is that the control rods are inserted from underneath the core. For this structural reason, its bottom section is 2–3 m longer than that of the PWR pressure vessel.

Another difference is the operating pressure. The BWR's operating pressure is approximately 7 MPa, which is about a half of that of the TMI's reactor (a 15MPa PWR). Because of its low operating pressure, the BWR pressure vessel can be built to have a large diameter.

One characteristic of BWR is that it contains more water in the reactor's pressure vessel compared to PWRs, particularly in the section underneath the core, not only because the BWR pressure vessel is larger in diameter and length, but also because it has guide tubes for sliding the control rods in and out underneath the core. How this characteristic contributed to the development of the core melt accident at Fukushima Daiichi is an interesting subject that will be discussed later on.

1.2   TMI Reactor's Performance During the Accident

Now then, on to the main issue. I will describe briefly, without getting into too much detail, the sequence of events that led to TMI ending up with the kind of core melt shown in Fig. 1.1.

First of all, PWR is composed of a closed-loop that circulates water by a primary coolant pump in order to transmit the heat generated by the reactor to a steam generator. The technical name of this closed-loop is the primary cooling system. Since water of such a high pressure as 15 MPa flows through it, it is also called as the reactor coolant pressure boundary.

The role of the circulating primary coolant (water) is to collect heat as thermal medium from the reactor and to deliver it to the steam generator. The steam used for driving a turbine generator is made from the water of a separate system that runs through the secondary side of the steam generator. This is how the PWR electricity generation works. It is different from the structure of a BWR in which the steam generated by the reactor is directly used for driving a turbine generator.

The feature of the PWR design is not to allow the primary coolant to boil. In order to accomplish this, it is equipped with a pressurizer for maintaining the primary cooling system's pressure to 15 MPa. The pressurizer is a tank with a capacity of approximately 40 m^3 and is placed at the highest position of the primary cooling system. Although I skip the explanation here, the TMI accident was caused by the relief valve located at the top of this pressurizer remaining open. The fact that they did not realize this problem led to the catastrophic accident.

In other words, there was an opening, i.e., the relief valve of the pressurizer, at the top of the primary cooling system in the high-temperature, high-pressure closed-loop. The primary coolant of course gushes out from this hole, which causes the water volume of the primary system to decrease. If the water volume decreases, the pressure drops and boiling occurs. The steam generated by the boiling gathers at the top of the pressure vessel, and starts to govern the pressure of the entire primary

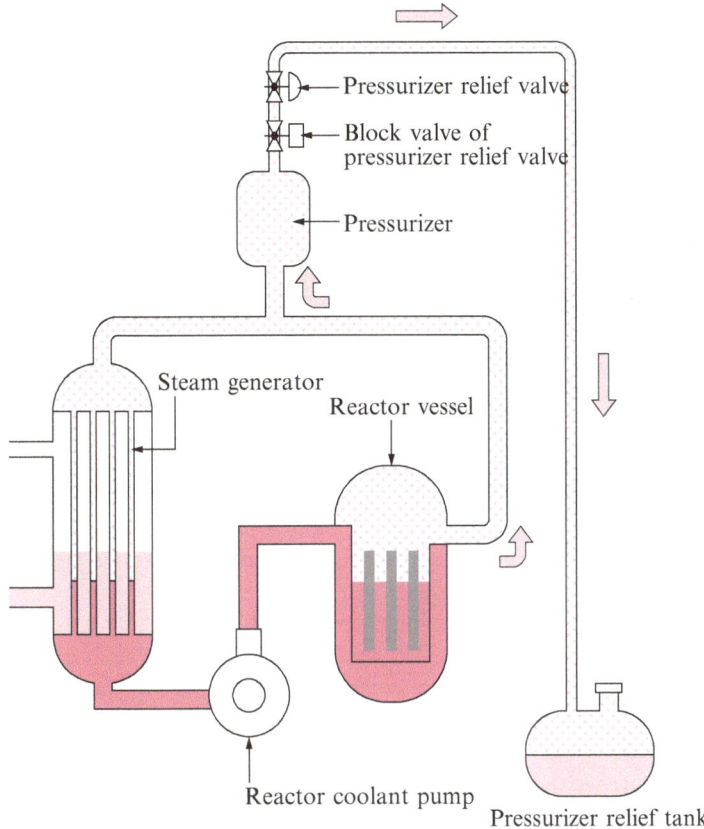

Fig. 1.3 TMI accident: reference diagram showing the status just before the block valve of pressurizer relief valve is closed

coolant system in place of the pressurizer. As the boiling starts, the pressure reduction speed drops, but the amount of the primary coolant keeps decreasing.

The water quantity of the core was dropped to about a half of what it should be just before the core melt took place (Fig. 1.3). The pressure of the primary system dropped to about 4 MPa as well just before the core melt. Until people finally realized that the pressurizer relief valve had been open and closed its block valve, the core was cooled by a mixture flow (i.e., two-phase flow) of water and steam.

The first effect of this was the vibration of the primary coolant pump. When steam enters a rotating pump, it causes a vibration phenomenon called cavitation in the pump. This cavitation vibration gets more violent as the steam volume increases, and ultimately become so violent as to break the pump. I personally experienced this while participating in the test operation of an elevated storage tank of the JRR-2 research nuclear reactor during the days I was working at The Japan Atomic Energy

Research Institute. It was such a tremendous vibration accompanied by a severe graunching noise that it shook the building that housed the pump. I remember that I ran out of the building thinking that it was a big earthquake. The vibration is that big.

Noticing the vibration, TMI's operators (staff) stopped the pump. It was approximately 100 min after the accident. All four pumps that constitute the primary coolant pumping system thus stopped. This was not a misoperation; you could say that the operators had no choice. However, if the pump stops, the water flow to the core stops and the fuel cooling effect will be lost. Naturally, this will cause the fuel rod temperature to rise.

Let us look at this change from the fuel rod side. As the influx of the coolant stops, so does its cooling effect, and the fuel rods are now simply immersed in a still water. Moreover, since the quantity of water has now reduced to about a half, only the bottom half of the fuel rods is soaked in the coolant. The top half of the fuel rods is exposed above the still water. The only consolation here is that the generated steam is flowing toward the pressurizer relief valve that is left open, so that the top half of the rods that is exposed can be cooled by said steam flow. In other words, the fuel rods at that stage were like bathers in an open-air spa with their upper bodies exposed to the wind. Putting aside whether they said "What a good bath!", the fuel rods were generally in a cooling state.

However, at 139 min after the initial mishap, the staff finally realized that the pressurizer relief valve had been open and closed its block valve. It is said that the person who noticed it was an operator-on-duty who came into work for his shift. Presumably he noticed it as part of the handoff procedure at the start of his shift.

Thus, an outlet of the steam was closed. With the outlet closed, the primary cooling system returns to the original state of a closed-loop. This means that there is no outlet for the reactor's heat. The decay heat causes the temperature and pressure of the reactor to rise. You could think of it as the breezy open-air spa turning into a hot and steamy sauna. The temperature of the fuel rods naturally rises. That was the start of the core melt.

The accident of the Fukushima Daiichi nuclear power plant had the same starting point. Please remember this change from an open-air spa to a sauna.

Now, let's take note of the term, "decay heat." If you know what it means, you may wish to skip ahead. Decay heat is the heat produced by the radioactivity stored in the fuel.

I am sure you are aware that the nuclear fuel generates heat as well as radioactivity because of nuclear fission reaction. I suppose you also know that there are three kinds of radiation, i.e., alpha (α) ray, beta ray (β) and gamma (γ) ray.

γ-rays are, for example, radiation with a very strong penetration capability just like the X-rays used in the medical check-ups. Having penetration capability means having energy. As a reverse process of generating the X-rays used in medical examinations from electricity, it is possible to convert X-rays or γ-rays into electric energy. The same can be said for α-rays and β-rays. All of them are energy in essence.

Table 1.1 Major events of TMI accidents

Key	Time after accident	
①	139 min	Closed the block valve of pressurizer relief valve
②	174 min	Turned on Reactor coolant pump (2B)
③	176 min	Declared site emergency
④	192 min	Opened block valve of pressurizer relief valve
⑤	193 min	Turned off Reactor coolant pump (2B)
⑥	200 min	Turned on high pressure water injection pump
⑦	224 min	Core melt material dropped to pressure vessel's lower plenum
⑧	Approx. 10 h	Hydrogen explosion in containment vessel
⑨	Approx. 16 h	Turned on primary coolant pump (1A)

When a nuclear reactor is operating, the radiation emanating from the radioactivity stored in the fuel rod turns into heat. This heat is called the decay heat. The decay heat amounts to approximately 7 % of the output of a nuclear reactor. Therefore, when the reactor ceases to operate, the heat generated by nuclear fission turns to zero, but the decay heat that amounts to 7 % of the reactor output remains. Although the radioactivity of a radioactive material decays at the rate determined from its half-life, it does not disappear soon. As a general rule of thumb, the size of the decay heat is 7 % immediately after operation is stopped, 2 % after 1 h, 0.5 % after a day, and 0.1 % after 1 year.

From this point on, a series of key events (Table 1.1) that led to the core melt of the TMI accident will be described. Please simply read it through at this point since the meaning of this series of events will be discussed in detail in Sect. 1.5 of this chapter.

① 139 min: The staff realized that the pressurizer relief valve had been open and closed its block valve. As the flow from the pressurizer relief valve stopped, the reactor pressure started to increase gradually.

② 174 min: One of the primary coolant pumps that had been stopped was turned back on. The operating time was approximately 19 min. Presumably they sought to stop the rise of the primary coolant pressure, and establish the natural circulation cooling.

However, as the pump was restarted, an abrupt change of the core neutron flux and a simultaneous sharp increase in pressure occurred. It is said that this pressure increase was 5.5 MPa within a time span of 2 min. As described later, this was when the core collapse and melt occurred.

③ 176 min: The radioactivity of the primary coolant increased sharply, making it clear that fuel destruction had occurred, so a site emergency was declared.

④ 192 min: As the pressure of the primary coolant system increased, the staff opened the block valve of the pressurizer relief valve again to release the reactor's pressure.

⑤ 193 min: The restarted primary coolant pump was turned off.

⑥ 200 min: Water was injected into the reactor by operating the high pressure injection pump. It is thought that the core was completely covered by water from this point on.

Thereafter, until approximately 10 h after the accident, the operators repeatedly attempted to establish cooling by natural circulation through operating the block valve of the pressurizer relief valve or operating the high pressure injection pump, but it seems that they failed to realize that what was preventing the natural circulation was the non-condensing hydrogen gas accumulated at the top of the pressure vessel.

⑦ Approximately 224 min: The crust that contained the molten core broke down, and approximately 20 t of molten material flowed down to the lower part of the reactor vessel. This did not cause any steam explosion.

Although it was originally reported that small explosions occurred at 225 and 230 min in the pressurizer relief tank inside the containment vessel, this is now understood to have been a breaking of the rupture disk of the tank. The pressurizer relief tank is a small tank for condensing the steam releasing from the pressurizer relief valve back into water. In any event, as either the tank or the rupture disk was destroyed, the gas that blew out from the pressurizer relief valve was now discharged into the containment vessel without resistance.

⑧ Approximately 10 h later, a big hydrogen explosion occurred inside the containment vessel, and caused a sharp pressure increase. This pressure increase caused the containment vessel spray to operate. However, since the PWR's containment vessel is a humongous vessel, the effect of the explosion was contained inside the containment vessel.

⑨ After 16 h from the start of the accident, the staff turned on again the primary coolant pump that had been turned off. Although they were not too confident of what they were doing, this switch-on action turned out to be a big success. As the big pump began to run, the core temperature started to drop sharply.

Of the nine events described above, ① and ② are the events that are directly caused the core melt. On the other hand, ③ through ⑨ are the behavior of the core immediately thereafter and the steps taken in reaction to it. I will now begin the explanation on these nine events, but to do so, I need readers to understand the PBF experiment and the oxidation reaction of the zircaloy cladding. To some of you, this may all be new to you and you may already be sick and tired of this, but please bear with me; this is not so difficult to understand as it sounds.

1.3 Fuel Behavior Test Simulating Power Cooling Mismatch Accident (PCM-1) at Power Burst Facility (PBF)

The core of a 1,000 MW class nuclear power station is a structure comprising 40–50,000 fuel rods disposed in a neat array. In order to understand the core melt, we must know how the fuel rods will melt. Lesson 1 of the secret of Fukushima Daiichi accident is to learn about the melting of the fuel rods.

The Power Burst Facility (PBF) test conducted at the Idaho National Engineering Laboratory (INEL, which is now called INL), in the late 1970s was an experiment of how fuel rods melt that was conducted at close to real-life conditions. A water flow loop that simulated operating conditions of power generating reactors was placed in a reactor produced specifically for the experiment. Then, test fuel was placed into the water flow loop to observe the melting conditions of the test fuel. One of the experiments they conducted was called the Power Cooling Mismatch (PCM) experiment, in which fuel rods were heated to exceed the cooling capacity in order to cause them to melt by force. I will now briefly describe this experiment based on the descriptions that can be found in the reference [1] at the chapter end.

When a metal wire is intensively heated, it will turn red and eventually burn out. Similar to that phenomenon, the water tubes of boilers often burned out in the past. Although the water tubes of boilers are designed to cause the water flowing inside the tubes to boil during normal operations, the temperatures of the tubes increase sharply if the boilers are heated too much or if the water flowing in the tubes decrease too much. This phenomenon is called "burnout."

The reason the tube temperature increases abruptly is that a thin steam film is formed on the inner wall of the tube due to lack of cooling. This steam film prevents direct contact between the water and the tube, thus causing the tube temperature to rise sharply, resulting in a burnout. This temperature rise may differ depending on the circumstance but it can be several hundred degrees Celsius to over 1,000 °C.

In the early days of the development of nuclear power generation, more than half a century ago, the melting of fuel rods was believed to be caused by burnouts. It was thought that, if the fuel rods generate heat in excess of the cooling capacity, a steam film would form around the fuel rods. The temperature of the fuel rods that has thus lost contact with water would rapidly increase, causing the fuel rods to eventually melt. This thought is generally correct, and it may still happen today in research reactors, where the fuel material is different from that of light water reactors. This phenomenon is normally called Power Cooling Mismatch (PCM).

Approximately ten PCM experiments were conducted at PBF. PCM-1 was the first of such experiments, in which, fuel rods with a length of approximately 90 cm were allowed to generate heat of five times that of the rated power, to see how burnouts occur. However, even after they were kept red hot at an average temperature of 1,500 °C or higher for 15 min, the fuel rods did not melt. According to calculations, the highest surface temperature must have reached approximately 2,000 °C. Approximately 80 % of the center regions of uranium dioxide (UO_2) pellets should have exceeded the melting point and melted. Although it was far beyond the burn out heat flux of the water tube boilers, had surely been reached, the fuel rods showed no sign of destruction.

Ultimately, after 8 min, a faint sign of radiation was detected in the loop. That was it. The fuel rods of light water reactors in these days do not burn out easily. As the only change was the detection of radioactivity, they gave up continuing it and turned off the nuclear reactor. It was then that an anomaly suddenly occurred. The flow volume of the coolant that was running through the fuel rod decreased and a

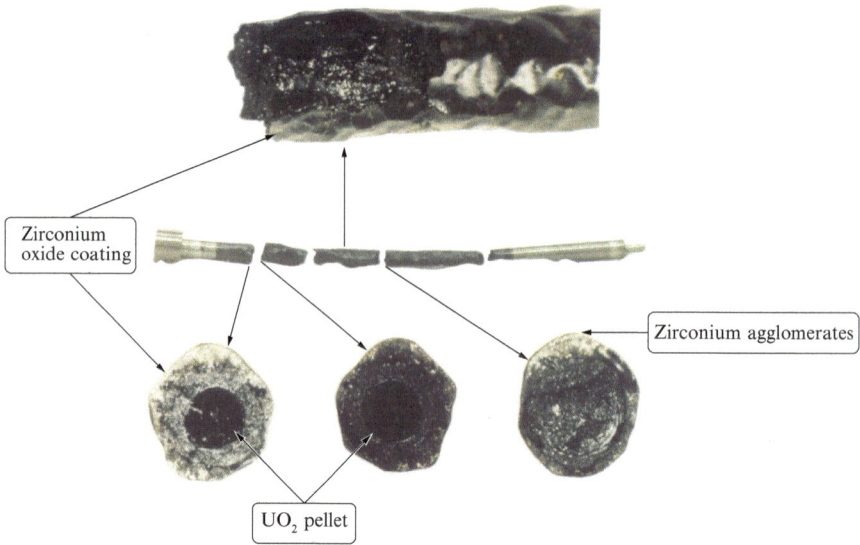

Fig. 1.4 Oxidated and split status of fuel rod in NSRR test

large amount of radioactivity was diffused in the loop. It was obvious that the fuel rod had been damaged.

"The fuel rod that withstood the intense heat broke down as soon as their temperatures dropped when the reactor operation was stopped. The tell-tale sign was radioactivity." It was a strange way of breaking.

When the test loop was opened, they found blackened and disintegrated pieces of fuel rods (see Fig. 1.4) laying over each other. The fuel pellets covered by blackened skin appeared to have various sizes and shapes of breaks and cracks, but they also seemed to have generally retained their original shapes. Although it was estimated by calculation that the inside of the pellets was in a molten state, it was reported that they saw no trace of the overlaid pellets welding together, in fact, there were some gaps formed between them for water to flow through. Although the flow volume of the loop decreased due to the increase of resistance caused by the piling of the decayed pellets, it also meant that the fuel was cooled by the coolant that flowed through the clearance.

Incidentally, this test result was a big news to be welcomed by us reactor safety researchers. It is a phenomenal fact that nuclear fuel does not burn out like boiler water tubes even an imbalance of output and cooling occurs, but rather breaks after cooling. "We can deal with this!" is what us researchers felt. This fact is useful for clarifying the Fukushima accident so that I will be referring to it repeatedly.

From the analytical test conducted later, a few important facts concerning fuel melt were found. The followings are the major findings from the test [1]. Since not all of them are directly relevant to clarification of the Fukushima accident, the readers may simply read through this for the time being.

The first finding is that the radioactivity release detected at 8 min after the test was caused by the oxidation of β-zircaloy in an approximately 10 cm long portion of the fuel cladding tube, where the temperature is the highest among the entire length of the fuel rod, the oxidation of Zircaloy progressed further due to the high temperature, zirconium turned into zirconium oxide, and the fuel rod broke. It is said that approximately 85 % of the inner portion of the fuel pellets were melted, leaving approximately 15 % on the outside intact, but no trace of dispersion of the molten fuel could be found in the coolant. The reason for this is believed to be that the zirconium oxide adhered to the solid materials served as an adhesive.

The second finding is the reason why the radioactivity release was small. It is said that, except the approximately 10 cm long portion in the middle of the fuel rod where the cladding tube was broken, the cladding tube softened by the hot temperature, pressed against the pellets because of the external pressure, and as a result formed a hermetic state. Consequently, it is considered that the radioactivity that existed in the gap between the cladding tube and the pellets was trapped in the cavities provided in the top and bottom end plug areas and was not released into the coolant.

The third finding is about the state of the top and bottom portions of the fuel rods where the temperatures were lower than that of the molten middle section. Of them, approximately 25 % were spilled out of the core in small pieces, while the rest of them were tightly bound by the film of zirconium oxide on the surface (Fig. 1.4), split into fragments, and dropped beneath the loop. The fuel in small pieces spilled out of the core after the test was finished, and no thermal disturbance such as steam explosions were detected during the test.

I have to add here that Fig. 1.4 is not a photograph from the PBF test but from the NSRR test conducted in Japan, which will be described later, to show how the oxide film adheres to the pellets [2].

The fourth finding is the fact that the zircaloy alloy of the cladding tube main body turned into agglomerates covered with an zirconium oxide film, and was found in various parts of the fuel rods. Most of them turned into zirconium dioxide (ZrO_2), which will be discussed later.

I suppose you must be feeling rather lost by now. But if you give up now you will not understand the core melt of the Fukushima accident. You will come to understand the secret if you read on. To tell you the truth, the researchers of the PCM test were also lost in the beginning.

The secret of the test is in the nature of the thin zirconium oxide film.

You may have heard about *yuba*, a Japanese food delicacy made from soy beans. It is essentially made from soya protein by thermally curdling it from a solution; it is dry and easily breakable when sold at the store, but when it is cooked turns into a sticky film that is pliable to wrap other foods with it. You may wish to imagine dry *yuba* as zirconium oxide in a low temperature, and cooked *yuba* as zirconium oxide at high temperature.

If you are not familiar with *yuba*, you could make analogous experiment by warming milk in a pot and observing what happens. Just before the milk comes to boil, it forms a thin film. It is surprisingly sticky, tough and unbreakable. Oxides are generally sticky and tough while they are at high temperatures but turn brittle when they are cooled. The oxide of zirconium used for the fuel rod cladding tubes very much have this nature. The coating gets thicker and forms a film. Hereafter, I'll call this simply as "coating."

It was after the test when the researchers latched on to this nature. One of the purposes of the PCM-1 test was to heat the fuel to a high temperature to see if it will cause the burnout phenomenon as in the case of boiler water tubes, but they found out that the nuclear fuel produces a tough oxide film on the surface as a result of the reaction between the cladding tube and water to prevent it from burnout. It maintains the shape of the fuel even after it turns red hot as the thin film stickily resists it from bursting.

This zirconium oxide film is very densely formed and does not easily allow water to penetrate through it once it is formed on the surface. Moreover, its melting point is close to 2,700 °C, which is much higher than the melting point, approximately 1,800 °C, of zirconium alloy (zircaloy), which forms the main part of the cladding tube. That is the key part of the secret.

The reason the cladding tubes did not burn out even though heat generation of the fuel rods was raised to 5 times the rated power and were left at intensely high temperatures for 8 min is that this thin film had a high melting point, was tough, and prevented contact between the hot fuel rods and water. The reason that a large amount of radioactivity did not spill out even when the cladding tube became brittle and finally broke was also because this dense oxide film tightly adhered to the pellet. The thin black film left on the surfaces of the disintegrated pellets found after the test are the remnants of this (Fig. 1.4). This is a very important finding for clarifying the core melt of TMI.

The broken down fuel pellets did not melt together, nor did they turn into agglomerates. They were overlaid on each other as if they were parts of a collapsed woodblock work, but there were spaces between them for the coolant to flow through.

The fuel rod was broken down into pieces because of the characteristics of oxides that become brittle when they get cold. The oxide film that was strong at high temperatures turned brittle as they got cooler due to the stoppage of the reactor. The oxide film that became brittle caused cracks as it shrank along with the temperature reduction. The tough films that clung to the fuel rod got torn apart here and there. It got torn apart mostly in between pellets. The fuel rod that no longer had the tough bondage of oxide films fell apart ("split") and collapsed.

As we can see here the fuel rod got split after it got cooler, so that broken fuel pieces never fused together when they fell on top of each other, thus leaving spaces between them for water to flow. This finding is also very important. Please remember this as well. These passages are known as communication paths.

Such are the descriptions and results of the PCM-1 fuel rod melt test. These findings are indispensable in considering the core melt in Fukushima. Let me summarize the key points.

The reactor fuel rods do not melt easily by simply applying heat. One reason for this is that the melting point of uranium dioxide is extremely high (2,880 °C) but the other is that the zircaloy used for fuel cladding tubes forms tough coating (with a melting point approximately 2,700 °C) as it gets oxidized. The coating gets brittle and breaks as soon as it gets cooler. Moreover, zirconium oxide serves as an adhesive by entering in between the molten uranium dioxide and maintains the shapes of fuel rods.

Fuel rods do not break down by disintegrating and melting when they are overheated as they are normally believed. Their oxide coatings break when they get cooler and split apart. The broken fuel pieces splitting apart, therefore, do not fuse with each other as they are already cooled on their surfaces. They further get cooled by water running between them.

I hope you understand that this is a decaying mode that we seldom encounter in general engineering practices, and that is because of the characteristics of the oxide coating of zircaloy which is used as the cladding tube material. The TMI core to be described in detail below reached its melting state after experiencing this process, but we need to study the oxide film of zircaloy in the next section to understand it fully.

1.4 Zircaloy Oxidation and Fuel Rod Behavior

We will first visit oxidation and reduction, which we learned in middle school. Zirconium starts to form a coat of zirconium oxide by acquiring oxygen through its reaction with water or water vapor when the temperature gets to about 800 °C. On the contrary, the reducing water loses oxygen to become hydrogen. The hydrogen thus created is the main culprit that caused the gigantic explosions at both in Fukushima Daiichi and Chernobyl that surprised the world.

The oxidation reaction of zirconium is vast in content and complex, but let me try to simplify it.

First, zirconium's oxidation reaction gets active as its temperature rises, while the thickness of the oxide film that is formed increases as the reaction speed increases. It is normally believed that the reaction becomes unstoppable when the temperature exceeds 1,300 °C. In other words, at above 1,300 °C, the reaction with water continues until all zirconium atoms get oxidized. The reason is that the reaction between zirconium and water is a very intensive heat generating reaction. As the reaction heat is an enormous 586 kJ per 1 mol, the heat generates more reactions, and more reactions generate more heat, and so on, i.e., a positive spiral phenomenon results.

Please allow me to digress for a moment. Safety review criteria are based on the results of the analyses of the Design Basis Accident (DBA). The judgment standard of the Loss of Coolant Accident (LOCA), one of the DBAs, specifies that the maximum temperature of fuel cladding tubes should be lower than 1,200 °C. This is a common rule worldwide. This particular setting is based on the finding of the oxidation reaction of cladding tubes that I mentioned above. The value of 1,200 °C was arrived at by subtracting a safety margin of 100 °C from 1,300 °C, and the purpose is to prevent a core melt that might be caused by the unstoppable oxidation reactions of cladding tubes.

Another explanation for the rule is that cladding tubes will be broken down due to thermal shock when they are quenched in reflood process of LOCA where cladding temperature had exceeded 1,200 °C. This is an explanation from the perspective of oxidation-caused brittleness of cladding tubes. Both are correct explanations.

Now earlier, we learned the PCM-1 test. We know that the fuel rods did not melt even though their surface temperatures exceeded 1,500 °C and that condition was maintained for as long as 8 min. On the other hand, safety standards say that an oxidation reaction would become unstoppable and the fuel rods would melt if the zirconium temperature exceeded 1,200 °C. Which is correct, you may wonder.

To be precise, both are correct. It is only that the former is an experiment using an actual fuel, while the latter is an experiment using zircaloy; different experiments may yield different answers.

Let's stop Zen riddle and talk straight. The finding that cladding tubes melt when they reach 1,300 °C is an answer obtained by causing zirconium, which was preheated to 1,300 °C or above, to react with water in order to obtain basic data. On the other hand, zircaloy cladding tubes of fuel rods used in reactors do not get to 1,300 °C all of a sudden. To reach 1,300 °C, there has to be a temperature history, e.g., 1 °C per 1 s. This makes the difference between the two cases.

The problem here is that, because the safety criteria of 1,200 °C has become so famous, surprisingly a large number of nuclear professionals believe that a core starts to melt when the surface temperatures of zircaloy cladding tubes exceed 1,200 °C.

We have studied the PCM test so we are now aware that this belief is quite incorrect. The actual fuel rods did not melt even after they reached 1,500 °C. They held quite firm. The reason was that the temperatures of fuel rods did not reach 1,500 °C all at once but rather each fuel rod had its own history of gradual temperature rise, during which an oxide film was formed on the surface of each fuel rod. Thanks to this oxide film, the zirconium in the interior of the cladding tube is protected from direct contact with water. Nonviolent reaction can occur because the oxide film is preventing it.

Now, please recall the TMI fuel rods just before the melt, which was discussed in the Sect. 1.2 of this chapter. It's the story about fuel rods soaking in a hot spring.

As they are, the bottom halves of the fuel rods are soaking in a still coolant bath. The top halves are being cooled by the steam flowing out from the pressurizer relief valve that remained open. Since it was approximately 100 min after the accident, the decay heat had dropped to about 1.3 % of the rated power. The reactor pressure had also dropped to approximately 4 MPa.

From the above status, it is safe to assume that the temperatures of the bottom half of the fuel rods were in the neighborhood of 250 °C, approximately equal to that of the coolant in which the fuel rods were soaked. You can say that the top halves of the fuel rods were cooled by breeze. Let us assume the temperatures of the highest temperature areas of the fuel rods to be around 500–600 °C. The average temperature must have been around 400 °C. Under such a core condition, nothing in particular occurs to the fuel rods. That is a quite healthy state, no different from normal operating condition.

The picture of the core at that instant is one of approximately 40,000 fuel rods sticking out from the water surface in a neat grid-like formation. Steam would have been flowing around at a leisurely pace, and a breeze would be blowing on the fuel rods – a peaceful sight.

① 139 min: The staff realized that the pressurizer relief valve had been open and closed its block valve.

At that instance, the breeze that was cooling the fuel rods was lost, and the open-air spa suddenly turned into a sauna. There was no change to the bottom halves of the fuel rods soaked in water, but the top halves were not only steamed by the sauna but were heated by their own decay heat amounting to a little over 1 %, so that the temperature started to rise at the rate of approximately 0.74 °C per second. That was the start of the core melt.

Let me describe the basic performance of the cladding tubes in a model case with this TMI core status as the starting point.

Please refer to Fig. 1.5. This is a picture showing the cladding tube temperature and the conditions of the fuel rods at the time of the TMI accident. The dotted line in the picture represents the water level of the core. The "cladding tube temperature" shown in the figure represents the surface temperature of the cladding tubes in the steam, not in the water. Since the decay heat 2 h after the stop of the reactor was

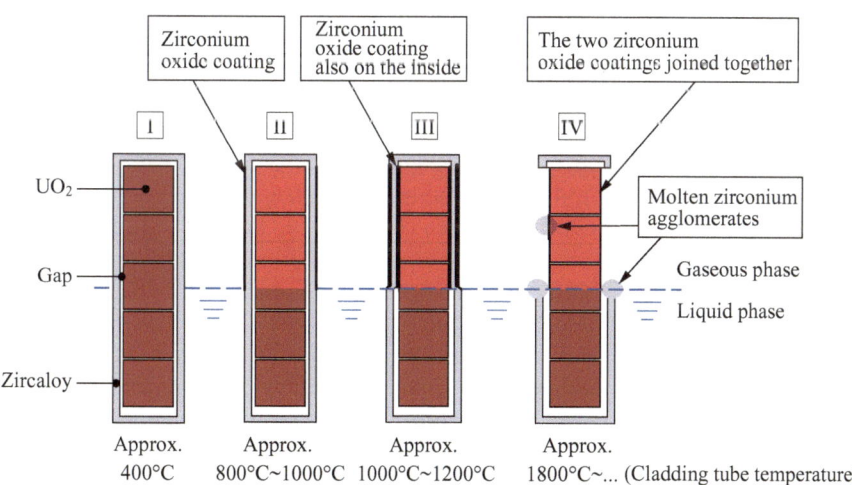

Fig. 1.5 Phase diagram of cladding tube temperature and fuel rod

a very minute heat generation that could barely increase the fuel rod temperature by 1 °C per second, please assume that the temperature of the fuel pellets was approximately equal to that of the cladding tubes.

Now let me move on to the explanation of the drawing.

[I] The state of fuel rod with the cladding tube temperature of 400 °C is the status of 100 min after the start of the TMI accident. The cladding tube is still not oxidized and the fuel pellets are still in a healthy condition. There is a gap between the pellets and the cladding tube where the gaseous radioactive materials that have leaked from the pellets dwell. This condition is not different from that of the fuel rods under the normal operating condition.

[II] When the cladding tube temperatures reach in a range of 800–1,000 °C (about 10 min after the temperatures start rising), the surfaces of the cladding tubes get covered by oxide coatings. While the oxide film looks black under the room temperature, it must have appeared red in the reactor. Although the cladding tube main body inside the oxide coating maintained its normal form, it must have started softening as the temperature rose. The melting point of the oxide film (ZrO2) is approximately 2,700 °C and is substantially higher than that of the melting point (1,740 °C) of zirconium that constitutes the main component of the cladding tube.

[III] When the cladding tube temperature reaches 1,000–1,200 °C (about 13 min after the start of the temperature rise), the tube itself becomes softer and starts to deform. This phenomenon was observed in the PCM-1 test as well. Since the fuel rods are under high pressure of the coolant, the softened tubes are pressed to the pellets to make close contact with them. When the inner surfaces of the tubes make close contacts with the pellets, the zircaloy on the inner surface reduces the uranium dioxide and produces an oxide coating [ref. to Appendix 1.1]. In other words, oxide coatings are formed on both the outer and inner surfaces of the cladding tubes so that zircaloy in the cladding tubes is in a condition sandwiched between the two oxide coatings.

When the temperature further rises in this state, not only the oxide coatings but also a three-element mixed melt substance (alloy) consisting of uranium, zirconium and oxygen is formed on the close contact areas in the interior of the cladding tubes are formed. Although no further explanation on this phenomenon will be provided here, I must say that the sandwich condition continues. In the following explanation of the fuel melting phenomenon, all of these will be discussed in a lump as a matter about oxides.

[IV] This is a picture depicting the fuel rods when the temperature rose further and exceeded the melting point of zirconium, i.e., 1,740 °C (about 25 min after the temperature started to rise). The zircaloy metal constituting the cladding tube main body which is the core of the sandwich here melts and flows down the clearance between the two oxide coatings to develop pools at various points. As a result, in those areas where zircaloy converges, the fuel rods will deform and have undulating shapes. On the contrary, in the areas from which zircaloy moved away, the thin outer and inner coatings contact with each other to form a single sheet of oxide coating and tightly compress the internal pellets. This phenomenon was also observed in the PCM-1 test. The coatings tightly wrap around the pellets just like wrapping plastic

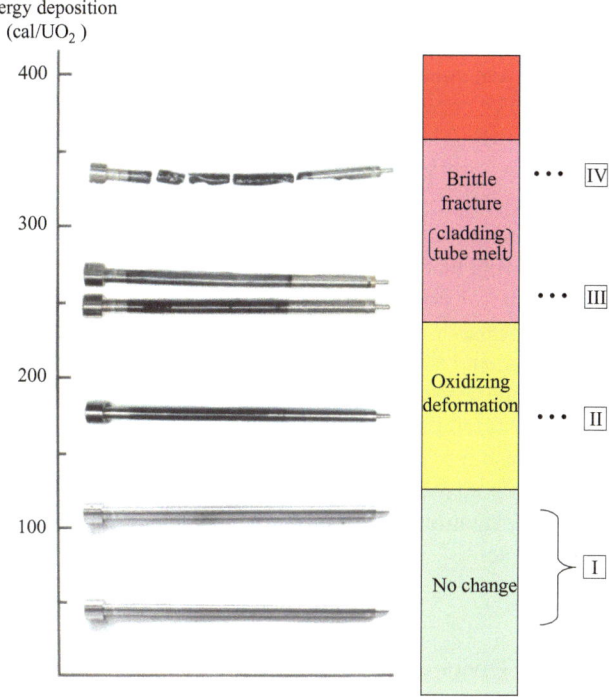

Fig. 1.6 Phase diagram of fuel rods in NSRR test result (Phase diagram number in Fig. 1.5)

sheets used to wrap around food in the kitchen, so that the fuel rods will remain standing up straight. The fuel rods remain standing straight up in the steam.

This [IV] fuel rod status itself was the status immediately prior to the core collapse.

You may think that I am telling you this as if I saw the status with my own eyes, but here is a test photograph that demonstrates that this model diagram is correct (Fig. 1.6). This is a photograph from the Nuclear Safety Research Reactor (NSRR) Test, namely the Fuel Test [2] concerning the Reactivity Initiated Accident (RIA), conducted by the former Japan Atomic Energy Research Institute (currently the Japan Atomic Energy Agency). Although a description of the NSRR Test will not be provided here, it was a research work, known all over the world, conducted in collaboration with the PBF tests since the latter half of 1970s, that clarified fuel performance during reactor accidents.

Figure 1.6 is a series of photographs that graphically documented various stages of the fuel as described as the steps [I] through [IV] in Fig. 1.5. Please look at these photos comparing them with the steps [I] through [IV] in the previous descriptions. You should be able to see for yourself that they are not the products of fantasies but the actual phenomena that can happen to fuel rods in reality.

Let us verify the model diagram in comparison with the PCM-1 test. Since the entire fuel rods were in a red hot condition during the first half, or 8 min, of the PCM-1 test, which was conducted for 15 min with 5 times of the rated power, it corresponds to the [IV] condition. The surface was covered by thin oxide coatings tightening up the pellets. The main body of the zircaloy cladding tube in between the coatings must have melted and converged in certain areas, so that the fuel rods which were once smooth cylinders must have become now irregular-shaped rods although still standing up. The zircaloy cladding tubes had already lost rigidity by that time but it is evident that the fuel rods of PCM-1 stayed standing straight up for 15 min because there was no report of change in the test data.

As soon as the test was finished and the heating was ended, the fuel rods became cooled by the coolant and the oxide coatings that became brittle developed cracks due to contraction and lost their tightening capabilities. As a result, the fuel rods came to split apart and fell into pieces ([IV] of Fig. 1.6), and radiation was detected in the coolant. It is also reported that the molten fuel in the middle fell off (from its original position) keeping the fuel rod shape due to the adhesive action of zirconium oxide. This is the explanation of the fuel damage in the PCM-1 test described in the Sect. 1.3 of this chapter.

1.5   Core Melt Process of TMI Accident

Now I will explain what happened in the TMI accident in a chronological order. Let us refer to Table 1.1 again.

① 139 min: The operating staff closed the block valve of the pressurizer relief valve and the breezy open-air spa turned into a sauna instantly. The top halves of the fuel rods that are exposed sticking out of the water surface start to develop oxide films form on the cladding tube surfaces as the temperature rises.

Just before 174 min – that is, before the operating staff restarts the primary coolant pump, – the reactor has been kept in a sauna-like condition for almost 25 min since the closing of the block valve of the pressurizer relief valve. The fuel rod temperature in the top half of the core must have been, in red-hot condition exceeding 1,500 °C. This is the same red-hot condition [IV] as in the PCM test. The fuel rods are still standing straight but their surfaces, which were once smooth and cylindrical, now have undulations here and there with zirconium lumps.

② 174 min: The operating staff restarts the primary coolant pump to inject a large amount of coolant into the core. The red-hot fuel rods are now splashed with cooling water.

③ 176 min: The radioactivity of the primary coolant increased sharply, as an indisputable evidence of the fuel damage, so that a site emergency order was declared.

I suppose that by now there is no need for explanation. The fuel rods were cooled abruptly, the oxidized cladding tubes became brittle, and the fuel rods once standing like forest trees fell apart ([IV] of Fig. 1.6). This is evidenced by a sharp change in the neutron flux. A disturbance of the neutron flux was recorded due to the deformation of the core.

What was inconvenienced by this collapse of the core was the bottom half of the fuel rods that have been kept healthy under water. This collapse caused a substantial impact.

Up to then, the decay heat has been transmitted to water, so that the fuel temperature was kept around 250 °C, the same as the coolant temperature. The all of a sudden, fuel rods debris dropped down from above on the fuel rods which had been kept healthy and piled upon them. For the lower portions of the fuel rods, it was literally a thunderbolt from out of the blue. The steam which had until then had escaped above was prevented from doing by the pile of fuel rod debris, so that the heat could no longer be removed from the water surface by boiling.

At this point, both the heat generated by the fuel debris at the top of the pile and the heat from the lower part of the fuel rods accumulate at the water surface. To make things worse, the red-hot zircaloy that used to form agglomerates here and there inside the cladding tubes can now flow out and contact with water because of the fuel rods split apart. Thus, red-hot zircaloy was now able to react directly with water. Furthermore, the amount of zirconium that could engage in this reaction was quite large a big problem. The large amount of water provided by the pump operation and the large amount of zircaloy made for a very active reaction at the liquid surface.

As a result of this reaction heat, the temperature of the debris and fuel rods in the vicinity of the water surface shot up and caused all of them to melt. Even the fuel cladding tubes at the lower part of the core, soaked in water, that had been kept healthy so far started to react as well. As explained before, this heat generation is large enough to melt uranium dioxide, which is the fuel of the reactor.

This is how the TMI core melt occurred. The melting occurred not in the red-hot upper portion of the core, but rather in the lower part of the core that remained in the water. The red-hot zircaloy agglomerates that flowed out as the result of the fuel rod collapse reacted with water to generate a large amount of heat, and created the TMI melt core as shown in Fig. 1.1.

It seems that the molten fuel agglomerates joined together, and first formed a thin egg-shell like coating as the area that was in contact with water got cooled. The coating then developed into a thicker shell covering the entire molten core as shown in Fig. 1.1. The bottom of the shell covered the molten core like the bottom of a pot and was supported from below by the bottom part of the fuel rods where the heat generation was low.

What is interesting is the top portion of the shell. Since the thickness of the shell did not change much while the molten core was growing, the shell which was still soft became flat by the weight of the debris on top of it, so that the molten core slid left (in the drawing) due to the weight of the debris (Fig. 1.1), made contact with the baffle plate made of a thin stainless steel piece surrounding the reactor, and made holes in the baffle plate by melting a part of it. It is said that about a half of the molten core was pushed out through this melted hole, and moved to the bottom spherical part of the pressure vessel.

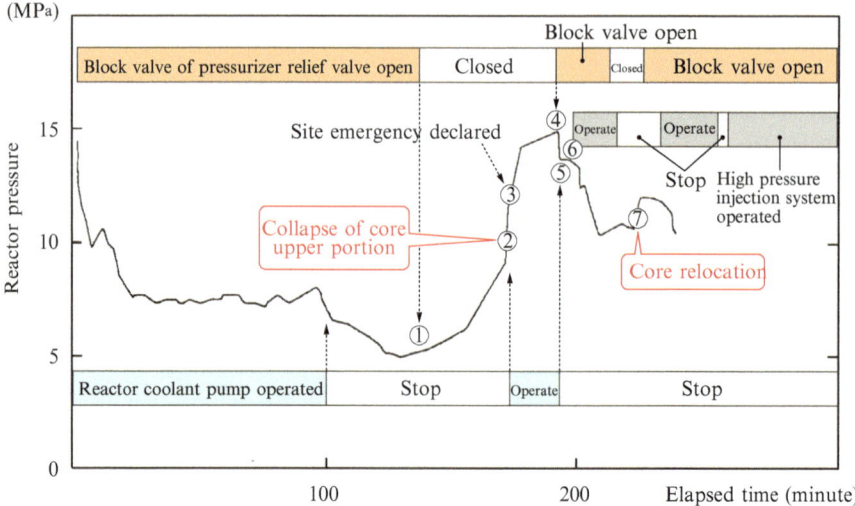

Fig. 1.7 Reactor pressure and accident sequence of TMI Unit 2 [3]

How long did the TMI core melt take? We can tell from the pressure change inside the reactor during the accident as shown in Fig. 1.7. At around 174 min, an abrupt pressure rise was detected. According to records, the pressure increased by 5.5 MPa within a span of approximately 2 min. That is a very steep increase in pressure. This kind of pressure increase could only be the result of hydrogen gas generation due to a metal-water reaction.

Let us make a ball-park calculation for a pressure increase of 5.5 MPa in 2 min. Let us estimate the total volume of the primary cooling system of the TMI reactor to be approximately 300 m^3, and assume that it contained 100 m^3 of saturated steam and 200 m^3 of saturated water. In order to achieve a pressure increase from 8.5 MPa to 14 MPa by zirconium-water reaction, it would take a heat quantity of approximately 4×10^7 kJ. In other words, it would require approximately 6 t of zirconium to react with water [Appendix 1.2].

In case of the TMI accident, it is calculated that a total of approximately 400 kg of hydrogen gas was generated. If we assume that approximately 6 t of zirconium were to react with water, that would generate about 260 kg of hydrogen, so our estimate seems by and large good. By the way, approximately 6 t of zirconium corresponds with approximately 25 % of the cladding tubes residing in the core.

The pressure rise during 2 min shown in Fig. 1.7 indicates that the core collapse and melting occurred about 2 min. As you can see, the core melt was an event of a quite short period of time. From this you can tell how violent the reaction was. Although the reaction between high temperature zirconium and water continued, it is believed that it was not as violent as the reaction that sparked the all-out core melt.

The core melt took only 2 min or so. The heat quantity that a zirconium-water reaction generates is that strong— that is the precious fact we learned from the TMI accident.

1.6 Settlement of TMI Accident

What happened after the core melt?

④ 192 min: The operating staff opened the block valve of the pressurizer relief valve. This was to release the reactor pressure which rose sharply by the generation of hydrogen gas.

⑤193 min: The reactor coolant pump was stopped.

⑥ 200 min: Water was injected by operating the high pressure injection pump. From this point on, the core was kept covered by water.

⑦ 224 min: The ever-growing molten core was pushed away by the weight of the debris accumulated on its top, contacted with and resultantly melted the baffle plate surrounding the core, and eventually flowed down to the bottom of the pressure vessel.

⑧ A large explosion occurred at 2:00 p.m., i.e., 10 h after the accident. I believe I heard from someone that the explosion occurred at the mezzanine level of the containment vessel.

It is a very important hint in analyzing the hydrogen explosion that this large explosion occurred 10 h after the accident. It gives us a hint that hydrogen gas does not explode until it meets an ignition source even after it has reached an explosive condition. In case of the TMI accident, it can be said that it waited approximately 7 h for the ignition source. Although we do not know what ignited the explosion 10 h later, it is still very useful for us in the analysis of what happened in the Fukushima accident to know that a hydrogen explosion does not occur without an ignition source.

Let's get back to the core melt point. When the reactor coolant pump was turned on, the neutron flux changed and the reactor pressure rose very sharply to an abnormal level. The radiation level rose at various points. The operating staff probably had no idea what was happening. I am sure the operating staff must have been confused and panicked.

Luckily, different from Fukushima, the lighting was on in the control room. The instrumentation and alarms were operating as well. The operating staff could grasp all the data they needed including the reactor pressure and temperature changes. And yet, they were not aware what caused the accident at that point. I am not criticizing them at all. I am simply trying to tell you that it is difficult to grasp the accident status even if it is in a situation where the lights are on and the instruments are working, entirely different from the case of Fukushima.

Let us now trace, for a while, how the operating staff operated then.

Although they tried to inject high pressure water after stopping the reactor coolant pump in order to establish the natural circulation (cooling by means of the secondary system), it didn't work. A large amount of hydrogen, a non-condensible gas, had been generated in the primary system and occupied the spherical dome in the upper portion of the pressure vessel, so this attempt failed. The operating staff suspected that the natural circulation of the primary system was obstructed by the presence of steam bubbles, so that they tried to break the steam bubbles through pressurizing the primary system by means of closing the block valve of the pressurizer relief valve. Next, they tried to lower the reactor pressure to the operating pressure of the residual heat removal system by opening the block valve of the pressurizer relief valve, but that didn't work either.

Additional injection of high pressure water did not allow them to add much water either as hydrogen gas worked as a cushion. On the other hand, if they stopped feeding water, the decay heat would increase the pressure so that they had to have kept the relief valve open. The operation staff kept opening and closing the block valve of the relief valve repeatedly.

It may sound silly to you who know the cause of it, but the operating staff was desperate as they did not understand what was happening. They repeated the endless operations again and again. I can imagine those operators repeating the same operations again and again with typical American cursing words. The sight comes to my mind without effort.

After about 16 h, they finally realized that there must be some kind of non-condensable gas, although they don't know what it is, that is occupying the reactor and preventing the circulation of the primary system. That was correct.

⑨ After 16 h of the start of the accident, the staff turned on again the reactor coolant pump that they turned off earlier.

Obviously they were not too confident of what they were doing. The report says "Turned on the pump just as a try." I am sure they were scared to turn on the switch. They must have done it so very tentatively. It was because they had to stop the pump due to the terrible cavitation vibration, and then when they turned it on again, the reactor pressure rose sharply with a violent change of neutron flux.

However, this very tentative switching on ended up, a big success. As the big pump began to run, the core temperature started to drop sharply. The circulation of the primary coolant was achieved when the pump began to run. I bet some smiles returned to the operators' faces and they started to make some jokes again. Like, "Hey, we did it!" I can almost see them jumping up and down.

At that point, they must have determined to run the pump all the way, no matter how much the pump would violate by cavitation. As they kept the pump running, the core cooled down. This is how the core shape survived as shown in Fig. 1.1 and that's why we can see it today. The condition of the core only 13 h after the core melt remains even today only because of this very tentative switch-on action. Without this picture of the TMI core melt, this book could have never been written.

Now, the status that the molten core is supported by the remains of the lower part of the fuel rods are attributed to the fact that the melting point of the supporting fuel (uranium dioxide) is higher than the melting point of the molten core. The reason that the melting point of the molten core is lower is that it consists of a three-element mixed melt substance (alloy) of uranium, zirconium, and oxygen with a melting point of approximately 2,000–2,200 °C. This fact had been known by a German experiment before the accident, and I heard that it was conveyed to the U.S. immediately after the TMI accident.

The TMI core melt illustration was completed circa 1989.

The most interesting is the existence of the debris remaining on the molten core. After the fuel rods that used to stand straight up collapsed when they were exposed to cooling water, the debris was cooled on the shell of the molten core. Instead of being rolled up inside the molten core, the debris stayed outside of it maintaining its shape. Without the existence of the debris, we cannot prove the relation between the PCM-1 test and the TMI accident. We would have never been able to explain the mechanism of the core melt as well.

Because the TMI core, which once melted was cooled within ten several hours after the accident, it was fossilized at the stage before it reached a complete melt to show us today how it was formed.

The TMI accident was essentially over when the reactor coolant pump was started 16 h later. The rest of the time was spent purging out the hydrogen remaining in the primary coolant and keeping the system cool by feeding water to the steam generator and the natural circulation of the primary system.

That is the end of the story of the core melt and solidification of the TMI accident.

1.7   Conclusion from TMI Accident

As the conclusion of this chapter, I will compile here the facts that will be needed in the analysis of the Fukushima accident.

I. The light water reactor fuel (core) does not collapse by melting even if the temperature rises to a red-hot condition due to imbalance of output and cooling. However, if fuel rods that are in red-hot condition are quenched, the fuel rods will get split up into pieces and the core will collapse.

The collapsed fuel's debris will be generally cooled by the water running around them so they will maintain their shapes. This performance is considered a characteristic of the LWR fuel, wherein zirconium is used as the cladding tube material.

II. If a sufficient amount of coolant is supplied, the fuel rods that are in a red-hot status react with water and cause a core melt. The time period for this process is relatively short. The reaction time that caused the TMI core melt was only 2 min or so. It is because the reaction between high temperature zirconium and water is extremely active and generates a great amount of heat.

III. A large amount of heat and hydrogen gas is generated with the core melt. Since the melting time is short, the generated heat and hydrogen gas will sharply increase the internal pressure of the pressure vessel.

IV. The explosion of hydrogen gas in the containment vessel occurred 10 h after the core melt. This is considered as a result of the fact that there was no ignition source. In other words, even if hydrogen gas has mixed with oxygen to become a combustible gas, it does not cause an explosion unless there is an ignition source or it reaches an ignition temperature.

V. It is considered that if the molten core contacts with water, it produces a strong shell within a relatively short period of time. This shell is probably formed as a molten mixture of the zirconium oxide and uranium dioxide, and prevents the molten core from contacting directly with water.

VI. The composition of the molten core is a mixture of uranium, zirconium and oxygen, and its melting point was roughly 2,000–2,200 °C in the case of the TMI accident (although it may vary with the mix ratio of the constituting elements), substantially lower than the melting point (2,880 °C) of the fuel itself (uranium dioxide).

Those were the findings about the core melt and hydrogen explosion learned from the PBF test as well as the TMI accident. The most important conclusion is that the cause of the TMI core melt was not the reactor's decay heat but rather violent chemical reactions between high-temperature zirconium and water. So far has been the preparatory study. Based on these conclusions I through VI, we will move on to the study of the core melt and explosion of the Fukushima accident in Chap. 2.

1.8   Influences and Excursuses of the TMI Accident

The TMI accident was essentially over within one day, but it was only in the third day after the first incident that it developed into such havoc as the evacuation of the residents. I will tell you now about several incidents that occurred during this period which might be of some interests to you. If you are in a hurry, you may wish to skip this section.

Before the happening of the TMI accident, a core melt was considered to occur at temperatures as high as 2,880 °C, i.e., the melting point of uranium dioxide, which is the core fuel. Therefore, people believed in such a farce that a steel pressure vessel (melting point: approximately 1,400 °C) can easily melt and a concrete floor (melting point: approximately 1,600 °C) can also be easily melted due to the decay heat of the molten core that drops through a hole made in the pressure vessel by melting. The fact that they even developed a computer analysis code for that purpose tells us how firmly they believed in the fiction.

The people's concern was crystallized into the famous movie "The China Syndrome." God sometimes plays a fancy trick. The release of the movie was just before the TMI accident so that it was perfect timing. It was considered the best movie of the time that predicted the TMI accident and it tipped the scale of public opinion in the United States completely toward the anti-nuclear movement. Contract after contract of nuclear power station construction was canceled. The nuclear power generation equipment manufacturers which once boasted that they have backlogs of 10 years before the accident started to fold the tents as they faced the cascade of order cancellations. It took almost 20 years even in the United States to recover from the shock and the public opinion to swing back to realize the need of nuclear power.

The graphical representation of the core melt of the Fukushima accident which was repeatedly broadcast over NHK TV, that I mentioned in my "Introduction" is exactly a kind of graphical representation of the concern people had. This is also the picture of a meltdown most of the people who have an interest in nuclear power have in mind even today. "Core melt equals full-melting (fluidization)" is the wrong notion still resides in the minds of people who have interests in nuclear power. The wrong understanding is proven by the molten core diagram shown in Fig. 1.1. Nobody realized the fact that the molten core remains in the pressure vessel at that time.

The angel of safety, if you will, taught us through the TMI accident that it is wrong to rely solely on the mechanical facilities for safety (Chap. 6 of Part II), and also left us the core melt diagram to hint to us that there is no need for extreme concerns of the likes of the China Syndrome by leaving us the core melt diagram.

As I said before, the TMI accident, which occurred at 4 a.m., March 28, was over actually within less than 24 h after it started. It is quite ironic that the commotion started only after the actual disturbance in the reactor was settled. It is the mass communication media that creates commotion, unrest, harmful rumors, and other dreadful things.

In case of the TMI accidents as well, no clear accident report was available for some time after the emergency declaration was issued about 3 h after the occurrence of the accident, allowing various speculations and confused information to fly around in the community, so that residents were kept uncertain and confused. It was no wonder as it was the world's first nuclear power station accident. Even the U.S. Nuclear Regulation Commission (NRC) itself was seemingly unaware that a core melt was occurring. There was a factor of complacency as well.

It was three days after the occurrence of the accident, more precisely in the afternoon of March 31, when Harold R. Denton, the Director of the Office of Nuclear Reactor Regulation at USNRC went to the power station with his staff to collect information. On the same day, the governor of the State of Pennsylvania issued an advisory statement telling the state residents to evacuate the area voluntarily if they were concerned about radiation effects due to the hydrogen explosion. However, the hydrogen explosion had already occurred at around 2 p.m. on March 28, approximately 10 h after the accident, and its effect had stayed within the containment vessel, so that the voluntary evacuation produced essentially no useful effect.

This advice for voluntary evacuation was essentially too late. It caused an unnecessary commotion because of panicked actions fanned by the anti-nukes and mass media circles over the already occurred and settled hydrogen explosion. Although the situation was entirely opposite, the same thing can be said for the evacuation of the Fukushima accident discussed in Chap. 4 of Part II. The evacuation order in case of Fukushima caused problems because it was too early.

This kind of unnecessary damage can occur from time to time in a major accident or calamity. Damage by harmful rumors is one of the worst kinds of problems that could happen associated with such an accident of calamity. I wonder if we should punish those who intentionally amplify the damage and effects of disasters as the offenders for pleasure of the worst kind.

On the other hand, the Pennsylvania governor's announcement about the voluntary evacuation during the TMI accident was cool. He declared that "those who evacuated voluntarily would be paid for the period as a special vacation." Since it is a paid special vacation, it goes without saying that all jovial Americans joined in it. It is said that more than a hundred thousands of people took advantage of the paid vacation and visited their friends and relatives living in far off lands. It may not be unrelated to this cool announcement that the nuclear power plant is not too unpopular in the area.

It was when I visited TMI about 6 months after the accident that I found T-shirts in a souvenir shop in front of the power station which showed a reactor hugging Mr. Denton saying "I love you!" The accident was not so unpopular among the local people either. Those who criticized the accident harshly were those anti-nuclear power intelligentsia living far off from TMI.

Director Denton is a good natured, honest person. He modestly says, "I didn't do anything particular. I just collected all the information I could get, examined them, and provided their contents to the mass media circles, that's all." In fact, what he did was to collect the information and made announcements and explained the accident based on accurate judgment, as he says. His was exactly the textbook performance of the risk manager. Because of this, the commotion sparked by the confusion of information was settled down within a few days.

Director Denton issued a statement that the critical situation of the power station was over on April 7, and by April 28, hydrogen was gone completely from the primary system. In other words, the accident essentially was settled.

That was the whole story of the TMI accident. You may be surprised by the fact that the TMI core melt occurred not in the upper portion of the core where fuel rods were red-hot, but rather in the lower portion where the fuel rods were immersed in and near the water surface and still intact. Is that quite unbelievable to you? However, that is exactly the process of the core melt that actually happened. Allow me to be repetitive, but the core melt diagram of Fig. 1.1 is the evidence.

People have believed up until now that the core melt is a process in which the temperatures of the fuel rods reach their melting point close to 3,000 °C and the fuel actually become fluid to flow down. It took about 10 years for us to find out that the TMI core melt occurred through a totally different process. This long period of time made us forget about the TMI accident.

In particular, the Japanese nuclear industry people were occupied more by paperwork rather than spending time on more critical matters and paid almost no attention to the results of the big accident 10 years ago. They did not study much either. Thirty some-odd years passed without much study, and then they faced the day of reckoning, March 11, 2011. This is evident from the fact that there was nobody among the leaders on the day of the accident who would compare the core melt at Fukushima Daiichi to that of TMI. For 2 months, Tokyo Electric Power Company (TEPCO), Nuclear and Industrial Safety Agency, and the government insisted that the core did not melt.

I am sure many of you have impressions that you have never thought of a core melt occurring by the heat of the zirconium and water reaction. It was probably because you are unfamiliar with the characteristics of the constituent materials of the fuel rods but also because your knowledge is limited to the activities in the temperature range of only up to 1,000 °C or so from your daily experience, and the performances of those materials in the temperature ranges of 2,000 °C are completely outside of your experience. If you think that it is an unbelievable interpretation, that is because it is outside of your expectation.

To begin with, an accident becomes a disaster because an unexpected thing occurs. An accident ends up as a disaster because what happened is not included in the existing computational model. If an accident is expected, it should be included in the computer as a possible scenario, so that an answer and countermeasure should have been prepared. An accident ended up as the disaster because something that was never expected in the computer occurred, and it is natural that the computer can only spit out wrong answers no matter how much the inputs are fine tuned. It is necessary to think about the physical phenomenon that led to the accident by sorting out and analyzing the accident process based on the accident data before trying to solve the problem relying on computers.

The core melt of Fukushima Daiichi was caused by the chemical reaction between water and zircaloy of the cladding tube that became hot same as in the case of the TMI accident. This reaction was very strong so long as there was enough water and it ended up in the core melt within only 2 min in case of the TMI accident. Then, how did the core melt proceed in case of the Fukushima accident? The key is in the fuel rod temperature and the amount of water injected. When these two conditions were met, the core would melt and generate a large amount of the hydrogen gas that causes the explosion develops. In Chap. 2, we will examine whether and how the condition was met.

There is one thing that I want you to be careful about before we proceed to Chap. 2. Although the word melt makes you think of a high temperature fluid, the core does not turn into low-viscosity fluid from the start of melting. Fuel materials that have become very hot and soft exist in a mixed and adhered state. Since the temperature is very high, as high as 2,000–2,200 °C, the fuel materials do not become fluid even after it reached the melting point because the heat is dissipated as radiation heat. The error of the video image of the core melt mentioned in the beginning lies in this part.

Now let us move into Chap. 2 taking note of this point.

Appendices

Appendix 1.1

The phase of the contact surface between the inner surface of the zirconium cladding tube and the surface of uranium dioxide pellets is extremely complicated from the professional standpoint (in the order of micrometer). That is where a lot of

complicated things occur, e.g., uranium entering the zircaloy layer, zirconium entering uranium, and oxygen entering the zircaloy layer to change into a substance
called α-zircaloy, or reduced uranium metal solving, etc. However, prior to the core
melt, the α-zircaloy layer existed behind this complex boundary surface (i.e., on the
zircaloy side). The melting point of α-zircaloy is higher than that of the zircaloy
alloy which constitutes the cladding tube main body, so that it also serves as the film
for forming a sandwich, just like the zirconium oxide coating on the outer surface
of the cladding tube. Using this result, I refer to them in this book collectively as an
"oxide coatings" in order to avoid a complex explanation.

Appendix 1.2

Ball-Park Calculation on the 5.5 MPa Pressure Rise

Let us assume that the total volume of the TMI's primary coolant system is approximately 300 m^3, and 200 m^3 of saturated water at 8.5 MPa and 100 m^3 of saturated
steam turned into L kilograms of saturated water of 14 MPa and S kilograms of
saturated vapor by heat generated by a zirconium-water reaction.

First, the zirconium-water reaction formula is:

$$Zr + 2H_2O \rightarrow ZrO_2 + 2H_2 + 586kJ / mol.$$

The atomic weight of zirconium is approximately 91.2 and 1 t of zirconium is
approximately 1.1×10^4 mol.

When 1 t of zirconium reacts with water, it generates a heat quantity of approximately 6.6×10^6 kJ and approximately 44 kg of hydrogen.

Next, as saturated water of 8.5 megapascal occupies a volume of 0.0014 m^3 per
kilogram, and saturated steam occupies 0.022 m^3, the water volume W that exists in
the primary coolant system is:

$$W = 200 / 0.0014 + 100 / 0.022 = 1.5 \times 10^5 \, kg$$

The total internal energy U (8.5 MPa) in the primary coolant system is:

$$U(8.5MPa) = 1,328.3[kJ / kg] \times 200 / 0.0014 + 2,564[kJ / kg] \times 100 / 0.022$$
$$= 2.0 \times 10^8 \, kJ$$

As saturated water of 14 MPa occupies a volume of 0.0016 m^3 per kilogram, and
the specific volume of saturated steam is 0.012 m^3 per kilogram,

$$0.0016 \times L + 0.012 \times S = 300$$
$$L + S = 1.5 \times 10^5$$

Hence, $L = 1.4 \times 10^5 \, kg, \ S = 1 \times 10^4 \, kg$

The total internal energy U (14 MPa) in the primary coolant system is:

$$U(14MPa) = 1,548.4[kJ/kg] \times 1.4 \times 10^5[kg] + 2,476.1[kJ/kg] \times 1 \times 10^4[kg]$$
$$= 2.4 \times 10^8 \, kJ$$

$$U(14MPa) - U(8.5MPa) = 4 \times 10^7 \, kJ$$

In order to provide this much energy by a zirconium-water reaction, approximately 6 t of zirconium needs to react with water.

Also, when 6 t of zirconium reacts with water, it generates approximately 260 kg of hydrogen.

In this evaluation, the pressure rise due to the generation of hydrogen gas is not considered so the zirconium-water reaction is overestimated. (The contribution of the hydrogen gas generation to the pressure rising is estimated to be about a half of that of the heat generation.)

While there are various estimations as to the amount of hydrogen gas generated during the TMI accident, it is generally agreed that it was about 400 kg, so that 260 kg as the amount of hydrogen generated at this point is more or less appropriate.

References

1. Cook B.A and others (1976) Behavior of a failed fuel rod during film boiling operation (PCM——1 test in the PBF). Trans ANS, Jan. 1976
2. Ishikawa M, Onishi Nobuaki (1986) Fuel damage performance appeared in the NSRR experiment. J Atom Energy Soc Jpn 28(5)3
3. Broughton J et al (1989) A scenario of the Three Mile Island unit 2 accident. Nucl. Tecnol 87(1989):34–53

Chapter 2
Fukushima Daiichi Unit 1, 2 and 3 Accidents

2.1 Outline of Fukushima Daiichi Nuclear Power Station

As of March 11, 2011, the day the Great East Japan Earthquake hit the area, there were five nuclear power plants comprising a total of 15 units installed along the Japanese coast facing the Pacific Ocean, ranging from the Tohoku to the Kanto areas. They are, from north to south, Higashidori Nuclear Power Station (hereinafter "NPS") of Tohoku-Electric Power Co., Inc. (1 unit) in Aomori Prefecture, Onagawa NPS of the same company (3 units), in Miyagi Prefecture, Fukushima Daiichi NPS of Tokyo Electric Power Co., Inc. (hereinafter "TEPCO") (6 units), in Fukushima Prefecture, Fukushima Daini NPS of the same company (4 units), also in Fukushima Prefecture, and Tokai Daini NPS of The Japan Atomic Power Company (1 unit), in Ibaraki Prefecture (Fig. 2.1).

Of these five nuclear power plants, four NPS, i.e., Onagawa, Fukushima Daiichi, Fukushima Daini, and Tokai Daini, reported damage caused by tsunami as a result of the Great Eastern Japan Earthquake. Of these, Onagawa and Fukushima Daini were still receiving power from external sources after the earthquake and tsunami. At Tokai Daini, although all external powers were lost, two emergency diesel generators were able to operate and provide electricity.

Except Fukushima Daiichi, all power plants had electric power to use. Or rather, simply because electricity was available, the operating staff's efforts were not wasted – they were able to stop the reactors, and they succeeded in safely cooling down the reactors. The damage brought by tsunami were wide and varied. The skill of the operating staff of all these power generating stations could only be judged as "commendable" for not causing any severe accidents despite various hazards. They proved that the claim of opponents to nuclear power that "human beings cannot control nuclear power" is absolutely wrong. People can control nuclear reactors so long as electric power is available. It is akin to saying that animals can live so long as water and food are available.

© Springer Japan 2015
M. Ishikawa, *A Study of the Fukushima Daiichi Nuclear Accident Process*,
DOI 10.1007/978-4-431-55543-8_2

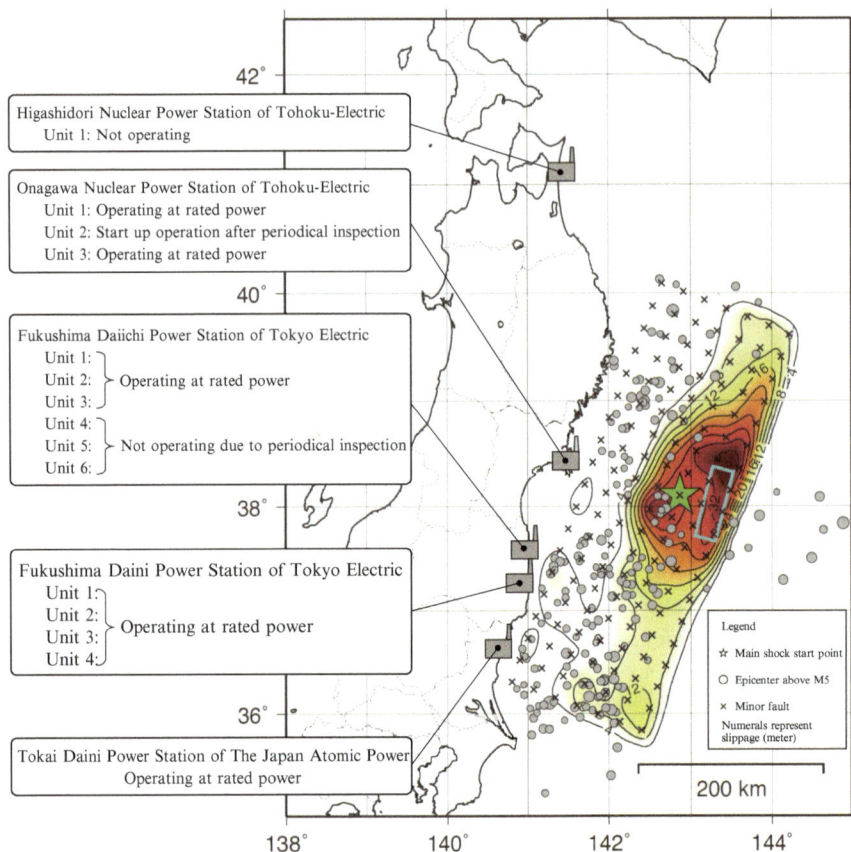

Fig. 2.1 Map showing relative locations of Great East Japan Earthquake and nuclear power stations on the Pacific coast of the Tohoku and Kanto areas (prior to earthquake) (Source: Meteorological data added with author's notes)

The tragedy occurred at Fukushima Daiichi, where electricity was lost. The difference from Onagawa, Fukushima Daini and Tokai Daini was that the external electricity supply was stopped for as long as 10 days due to the tsunami. Without electricity, the extraordinary effort of the power station's staff was unfortunately no more than a grasshopper's horn, a desperate, only symbolic attempt. Core melts and explosions occurred at four of the six reactors of Fukushima Daiichi, causing emergency evacuations of residents in the vicinity due to the fear of radiation.

I will attempt technical verifications and descriptions of how the Fukushima accident, particularly the core melts and hydrogen explosions, occurred.

Fukushima Daiichi NPS is situated approximately 220 km north-northeast of Tokyo covering portions of the towns of Ookuma and Futaba in the middle of Fukushima Prefecture's stretch along the Pacific coast, where the mainland of Japan, Honshu, is arching out its belly toward the Pacific Ocean accompanied with soft continuing hills. The plant is located on a hill of about 35 m above the sea level, and occupies an area of approximately 3.5 million square meters, extending about 3 km north-to-south and about 1.5 km east-to-west in the shape of a semi-circle. A pair of breakwaters extends in a triangular shape reaching approximately 0.7 km from the shore, and a dock and a wharf are provided inside the area guarded by the breakwaters.

Of the six reactors, Units 1–4 are located on the south side of the plant site, placed in the order of Units 4, 3, 2 and 1 from the south. Units 5 and 6 are located slightly apart from them on the north side, placed in the order of Units 5 and 6 from the south (Fig. 2.2).

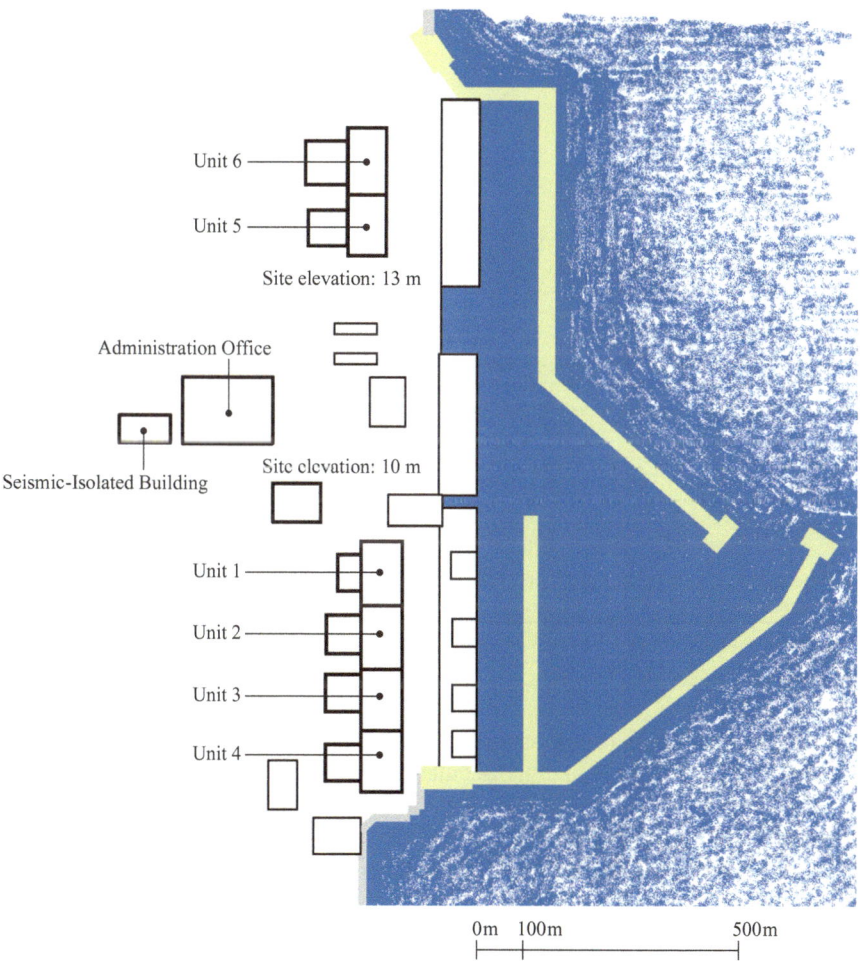

Fig. 2.2 Fukushima Daiichi NPS layout (Source: TEPCO data, with author's notes)

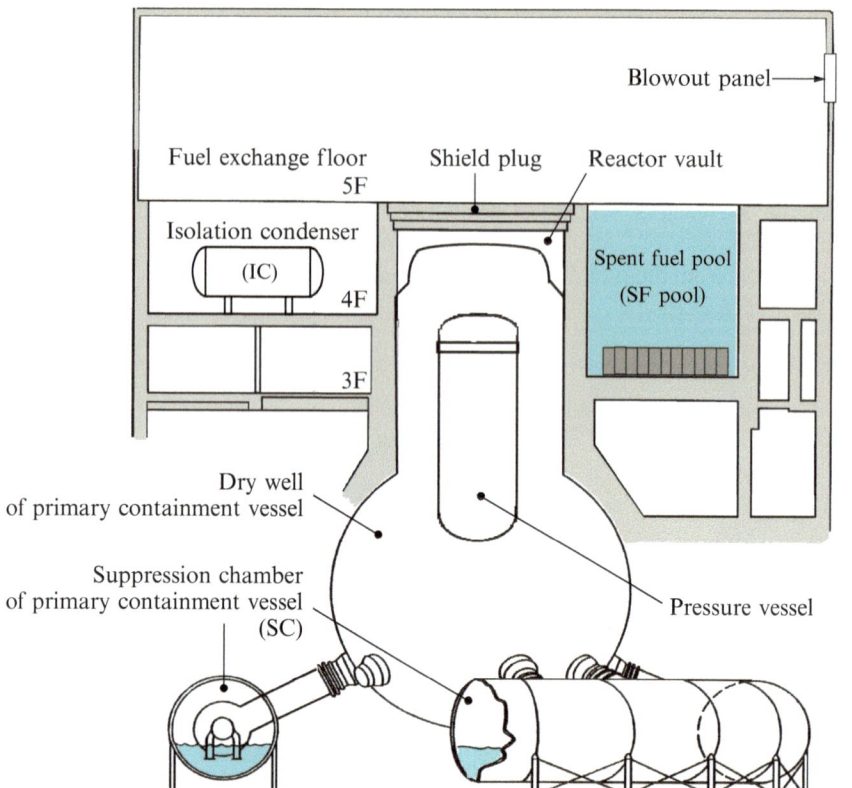

Fig. 2.3 MARK-I primary containment vessel (Source: from TEPCO's "Fukushima Nuclear Accident Report")

   The electrical output of Unit 1 is 460 MW while those of Units 2–5 were all 784 MW each. Their reactors are Boiling Water Reactors (BWR) and are each installed in a containment vessel equipped with a pressure suppression pool (or suppression chamber/"SC"), which is essentially a donut-shaped water pool, named MARK I (Fig. 2.3). Unit 6 is a 1,100 MW BWR and its containment vessel is a MARK II type, different from others, but the explanation of its detail will be omitted because it has no bearing to the accident.

   The major buildings (reactor building, turbine building, etc.) of the power station are located at 10 m above sea level in the case of Units 1–4, and at 13 m in the case of Units 5 and 6. They are built on the ground produced by cutting the soil of a hill of about 35 m above sea level down to a lower level on purpose. As there are various criticisms regarding the selection of the height of the power plant site above sea level, I will explain the reasoning for it in Sect. 5.3. The sea water pumps that played an important role in the cooling of the reactors were all located close to the shore and placed at 4 m above the sea level. As a consequence, they were all destroyed by the tsunami.

So that is the outline of the power station that seems to be necessary in studying the Fukushima accident.

2.2 Initiation of the Accidents – Earthquake and Tsunami

When the earthquake occurred on March 11, 2011, three out of the six reactors, i.e., Units 1–3, were operating while Units 4–6 were not operating because of periodic inspections.

An earthquake with a wide epicentral area, extending from off the coast of Iwate Prefecture to off the coast of Ibaraki Prefecture, occurred at 2:46 p.m. to initiate the Great East Japan Earthquake. According to a seismological record, the earthquake, which had a magnitude of 9.0, spread from the epicenter in the Sanriku coast at 2:46 p.m. to the entire sea area mentioned above, followed by numerous tremors including 8 earthquakes of level 5 or above on the Japanese seismological scale during a period of approximately 40 min until 3:25 p.m. (Table 2.1). It was a huge earthquake, without precedent in official records.

Because of this earthquake, all seven feeder lines to Fukushima Daiichi became unusable due to damaged breakers and collapsed transmission towers. In layman's terms, the power station suffered a power outage due to damage to the transmission wires. In professional terms, it is called external power loss.

The earthquake induced a big tsunami that hit a broad coastal area extending from Aomori to Ibaraki. About 45 min after the earthquake, i.e., at around 3:27 p.m., the first wave, approximately 4 m high, hit Fukushima Daiichi, and the second wave arrived at around 3:35 p.m. This second wave was so high that it well exceeded the tide gauge; according to a TEPCO estimation, it was approximately 13 m high. Units 1–4 were all covered by tsunami.

Although our common image of tsunami is that it approaches us at a great speed, scientists tell us that it is more like a high tide with a very deep expanse. Therefore, once it arrives, it stays in the same place for a while. In other words, the damage caused by tsunami is not caused by wave actions of water but rather by immersion in a pool of dammed-up water. Thus, a building hit by tsunami will be completely filled with water and mud, so that the mechanical equipment inside the building will become completely useless once they are engulfed by tsunami.

Most of the mechanical equipment installed in the basement as well as on the first floor of the power station became fully immersed in water by tsunami and became useless. We can see how extensive the tsunami was from the fact that most of the emergency diesel generators, more precisely 12 out of 13 diesel generators, provided for use in case of an external power outage, became unusable because the generators or related devices were covered by water. It was not just that. Since the power distribution panels were also installed either in the basement or on the first floor, most of them got immersed in water and became unusable. This meant that

Table 2.1 Earthquake and aftershocks in the Tohoku area of the Pacific Ocean

No.		Time of occurrence					Epicenter location	Dep	Mag	Maximum seismic intensity[a]
		Yr	Mon	D	H	Min				
1	Main shock	2011	3	11	14	46	Sanriku coast	24	9	7
2		2011	3	11	14	51	Fukushima coast	33	6.8	5 Lower
3		2011	3	11	14	54	Fukushima coast	34	6.1	5 Lower
4		2011	3	11	14	58	Fukushima coast	35	6.6	5 Lower
5		2011	3	11	15	6	Iwate coast	29	6.5	5 Lower
6		2011	3	11	15	7	Ibaraki coast	20	6.5	4
7	(Outside area)	2011	3	11	15	8	Izu (Shizuoka)	6	4.6	5 Lower
8		2011	3	11	15	8	Iwate coast	32	7.4	5 Lower
9		2011	3	11	15	12	Fukushima coast	39	6.7	5 Lower
10	Max aftershock	2011	3	11	15	15	Ibaraki coast	43	7.6	6 Upper
11		2011	3	11	15	18	Ibaraki coast	41	4.7	5 Lower
12		2011	3	11	15	25	Sanriku coast	11	7.5	4
13		2011	3	11	15	29	Sanriku coast	15	6.9	3
14		2011	3	11	15	59	Fukushima coast	50	6.8	3
15		2011	3	11	16	14	Ibaraki coast	25	6.8	4
16		2011	3	11	16	17	Fukushima coast	20	6.5	4
17		2011	3	11	16	28	Iwate coast	17	6.6	5 Upper
18		2011	3	11	16	30	Fukushima coast	27	5.9	5 Lower
19		2011	3	11	17	12	Ibaraki coast	32	6.6	4
20		2011	3	11	17	15	Fukushima coast	32	6.5	3

Source: Extracted from Meteorological Bureau's website
[a]Figures indicate those of Japanese seismic intensity scale (7 is the maximum scale)

even if the external electric power was restored, it was not easy to run the machines. As you can see the tsunami's strike on the electrical facilities was a double punch. The hardship of the operating staff started with this hard reality.

Now, back to Units 1–4. In the end, in addition to the loss of external electric power supply, the emergency power supply and DC (battery) power supply were also lost, so that the power station fell into a real emergency, losing all electricity supply. It's not just the machines that stopped to operate. All indicators stop to provide their readings, warning signals, and even lighting in the operation control room was lost. The control rooms in the windowless buildings of the nuclear power station were pitch dark.

This total power outage continued for approximately 10 days, until a temporary power system was installed on or around March 20 to restore electricity. As a result, the condition of the power plant slowly began to be restored. The restoration of electricity finally allowed the power station to turn the corner of the emergency.

In essence, the accident occurred when they lost electricity and, with the return of electricity, they began to see the light ahead. All the key subject matters of this book, including the core melt, explosions and emission of radioactive substances, happened during the dark period when no electricity was available. This makes us realize that the first cause of the accident at Fukushima Daiichi NPS which led to the disaster was the tsunami, but the second cause was that the total power outage lasted for 10 days.

The total power outage for an extended period of time was a main cause – one that is as significant as the earthquake and the tsunami.

Column Brilliant Performance of Fukushima Daiichi's Operating Staff
Digressing a little here, I wish to mention that there was one emergency diesel generator that survived by a stroke of luck. The operating staff who controlled Units 5 and 6 that day used this sole surviving generator to maintain the two nuclear reactors at cold shutdown by managing the scarce electricity available from this sole generator. I believe that this was indeed a commendable maneuver, because Units 5 and 6 were still generating decay heat although they were not operating, as they were under periodic inspections when the accident occurred. With this fact alone, it is clear that the operating skills of Fukushima Daiichi's operating staff were on par with those of Onagawa and Fukushima Daini – at world-class level.

This fact is well noted by the IAEA's investigation team who visited Fukushima after the accident and reported by the overseas mass communication services. Strangely enough, the Japanese news media failed to report that the accident could have been prevented only if the electric power supply was available, even though the power station was hit by the tsunami.

Moreover, the extraordinary effort of Fukushima Daiichi's staff was not appreciated very much as the Kan administration branded TEPCO as being responsible for the accident. This was despite the fact that the operating staff who gallantly stayed to battle the situation at the site under further deteriorating condition after the explosion of Unit 4 were hailed as the "Fukushima 50" by reporters overseas. As a witness to history, I would like to memorialize permanently this successful and gallant fight of the staff of Fukushima Daiichi.

2.3 Outline of the Accident

First, let me give you the outline of the Fukushima accident (Table 2.2).

On the morning of March 12, the core of Unit 1 melted, and the hydrogen explosion occurred in the reactor building at around 3:30 p.m. It was the first step of the disaster. With this explosion, we lost all hope of cooling down Unit 2. The cable prepared for connecting a power supply-car to the power distribution panel and

Table 2.2 Summary of the Fukushima Daiichi accident

	Unit 1	Unit 2	Unit 3	Unit 4
Earthquake attack	Mar/11 2:46 p.m.			
Tsunami attack	Mar/11 3:35 p.m.			
Core melt	Mar/12 about 4 a.m.	Mar/14 about 10 p.m.	Mar/14 about 10 a.m.	
Damage timing (structural)	Mar/12 3:36 p.m. (reactor building)		Mar/14 11:01 a.m. (reactor building)	Mar/15 6:14 a.m. (reactor building)

luckily survived the tsunami in order to feed electricity to a high pressure injection pump was now damaged by the explosion. This made it difficult to restore the electric power needed for cooling.

The reactor core of Unit 3 also started to collapse around 1:00 a.m. on March 13, and on March 14, hydrogen explosion occurred in its building. As a result of the core melt of Unit 3, i.e., because of the hydrogen gas flow from Unit 3, an explosion occurred in Unit 4 as well on March 15.

Around 10 p.m. on March 14, the core of Unit 2 melted, but barely escaped an explosion. This was because the explosion of Unit 1 blew off the blowout panel of the reactor building of Unit 2 and allowed the hydrogen gas to be exhausted. On the other hand, the radioactive substances that leaked out directly from the containment vessel severely contaminated the surrounding environment.

Speaking of radiation, the radiation level measured in the vicinity of the main gate fluctuated quite significantly and looks complicated at first glance, as shown in Fig. 2.4. However, by a closer observation of the graph, it can be seen that the background radiation dose increased with the core melt, and also some irregular increases can be observed occurring quickly due to vent operations, etc. With a more precise analysis, one can see further that the background radiation level increased twice.

The first radiation level rate increase that was caused around 4 p.m., March 12 was due to the core melt in Unit 1. The venting in Unit 1 occurred around 10 a.m., March 12, so that said increase has nothing to do with the venting. It seems that the radioactive substances which had been entrapped in the fuel rods were directly discharged due to the core melt from the building although it was only a minute amount. As a result, approximately 4 μSv/h of radiation were recorded in the vicinity of the front gate of the power station. This is not a radiation level that mandates an evacuation of residents. This condition continued until March 15 despite the emission of radioactive substances due to other venting events or the explosion in Unit 3.

However, the direct emission of radioactive substances started with the damage of the containment vessel of Unit 2 on the morning of March 15. Because of this direct emission, the radiation dose rate in the vicinity of the front gate jumped up almost 100 times to approximately 300 μSv/h (Fig. 2.4). This is a radiation level that mandates an evacuation of residents.

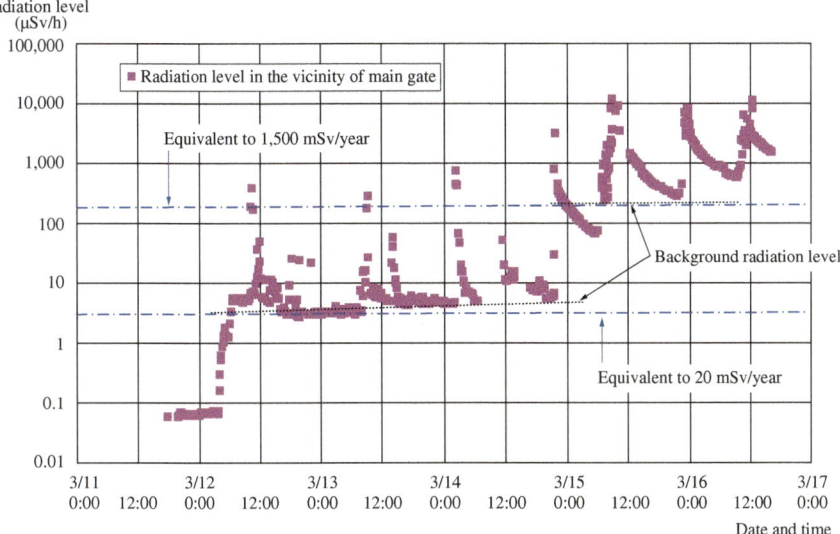

Fig. 2.4 Radiation level change in the vicinity of the front gate of Fukushima Daiichi NPS (Source: from TEPCO's "Fukushima Nuclear Accident Report")

However, the actual evacuation had already started as early as midnight of March 11. It was an emergency evacuation and was not based on any evacuation plan or preparation. The fact that SPEEDI (System for Prediction of Environmental Emergency Dose Information) was not implemented at that time shows well how haphazardly the evacuation was done as was criticized by the mass media. This lack of preparation for evacuation not only caused confusion and frustration among the evacuees but also caused up to 60 people to die at seven hospitals and long-term care facilities related to the evacuation [1]. I will discuss the evacuation issue again in Chap. 4 of Part II in relation to the radiation emission issue.

One other thing we must not forget is the extravagant operation for cooling of the spent fuel pool ("SF" pool) of Unit 4, i.e., the aerial water spray by the JSDF's heli- copter, the use of the Tokyo Fire Department's fire trucks designed for high-rise building fires, etc. That was an incident separate from the reactor accidents, more specifically the core melt and the hydrogen explosion, that are the objects of this book.

By March 20, temporary power was made available to the accident site. The restoration of lights to the once pitch dark accident site came to accelerate the han- dling of accident. The decay heat that caused the core melt and hydrogen explosion dropped down to well below 1 %. So long as the cooling of the core continued, the possibility of new unexpected surprises would be lower. There was finally some breathing room. As electricity became available and the water used to cool the core melt changed from sea water to fresh water, a new concern was what to do with the waste water discharged from the core melt and accumulated in the basement of the reactor facility. It was April 5 when we received complaints from South Korea and other neighboring nations for releasing low level contaminated water.

At this point, the difficulty the power plant staff was facing was enormous. Several hundred staff members were living as a group in a contingency planning building called "Seismic Isolated Building" that withstood the earthquake and tsunami. They slept on the floor with no change of clothes, and worked around the clock they were not quite prepared for. The food they ate was all cold preserved food.

And yet, they endured. I believe that those who worked at the power station, were the ones who recognized the graveness of the accident most and were most concerned about it. Most of them lived in the same community and stayed at the power station although their families had evacuated. During that period of time, almost no communication was possible between them and their families. It was approximately 2 months after the incident that signs of improvement came to be seen when the government and TEPCO jointly issued the first amendment to the "Roadmap towards Restoration from the Accident at Fukushima Daiichi Nuclear Power Station (May 17, 2011)."

If I were to write the story of all the difficulties the staff went through, it would easily fill a separate book, but it will deviate from the purpose of this book, so I won't say anything further. The IAEA's accident investigation team, which visited the site on May 24, offered the staff praise: "The on-site response by dedicated, determined and expert staff, under extremely arduous conditions, has been exemplary and resulted in the best approach to securing safety given the exceptional circumstances." As those who are familiar with power stations, they no doubt could not avoid making such an observation.

By June, with the help of U.S. and France, a recirculating cooling facility was completed on the site. The pipeline facility for removing the radioactive substances from the accumulated waste water so that the water can be recycled as cooling water was finally completed in a rush. With the completion of this facility, it became readily possible to cool the reactor and the temperatures surrounding the power plant began to drop gradually. As a result, the amount of radioactivity released from Units 1–3 dropped substantially, and it is reduced to about one one-hundred millionths of what it used to be at the time of the accident now 3 years later.

I suppose that it was after the interim report of the government on the accident, which came out around the spring of 2012, or about a year after the accident, when the decommissioning issue came to be talked about. Concerning the TEPCO's decommissioning plan, which took 4 years to formulate and will take 40 years to complete, I have been asked what I think of it by various people in the media as I have an experience of decommissioning the JPDR (Japan Power Demonstration Reactor) of the Japan Atomic Energy Research Institute about 20 years ago. I will be rendering my opinion about it in Part II, Chap. 7, so that I will stop my outline description of the accident right here.

2.4   Case of Unit 1

Unit 1 was completed and started to operate in March 1971. It was the third BWR built in Japan. It was the time when there was a boom in the construction of nuclear power stations in the United States, the world leader in the field of nuclear power generation, and most of the design and manufacture of their key components were done by General Electric (GE).

JPDR of Japan Atomic Energy Research Institute which succeeded in atomic power generation for the first time in Japan was also made by GE in 1963. The second oldest in Japan was Unit 1 of Tsuruga NPS of The Japan Atomic Power Company, which [was aimed for commercial atomic power generation for the first time] in Japan. That was also made by GE. The design of Unit 1 of Fukushima Daiichi is almost identical to those of other BWRs in use today but its electric output is approximately 460 MW, which is slightly smaller than the others.

If I were to point out the characteristics of Unit 1, it relies more on mechanical devices, as often seen in old machines. In that sense, it is a power station born in the midst of evolution. Comparing it to newer power stations is like comparing a manual shift automobile to an automatic shift automobile of today. However, it has no essential difference from the newer models in the basic performance, as manual shift automobiles are to automatic shift models. I see not a few operators feel attached to Unit 1 as if they love manual shift cars.

I wanted to write this because there are people who claim in know-it-all attitudes that Unit 1 is 40 years old and that's why it caused the accident. The accident was caused by tsunami and has nothing to do with the age of the power station.

2.4.1   Isolation Condenser (IC) Problem

Another feature of Unit 1 is that it has isolation condenser (ICs). The ICs, which are criticized because they did not work in the accident, are emergency cooling devices utilizing the natural circulation system used only for three older BWRs including Fukushima Daiichi Unit 1.

The Reactor Core Isolation Cooling System ("RCIC") is used instead of ICs for all units after Unit 2. In that sense, Unit 1 is an old type of nuclear reactor. Although I will be explaining about RCIC in relation to Unit 2 later, please remember for the time being that both ICs and RCIC are safety devices for the purpose of cooling the core when the reactor is separated (isolated) from the turbine as in the case of this accident.

Let me make here a simple explanation of ICs. As shown in Fig. 2.5, the essence of an IC is a simple heat exchanger for cooling purposes, which is located at a position higher than the reactor. It is a cooling circuit based on the principle of natural circulation in that the steam rising from the reactor is cooled and condensed to return to water as it passes through the heat exchange tube. It is a highly reliable

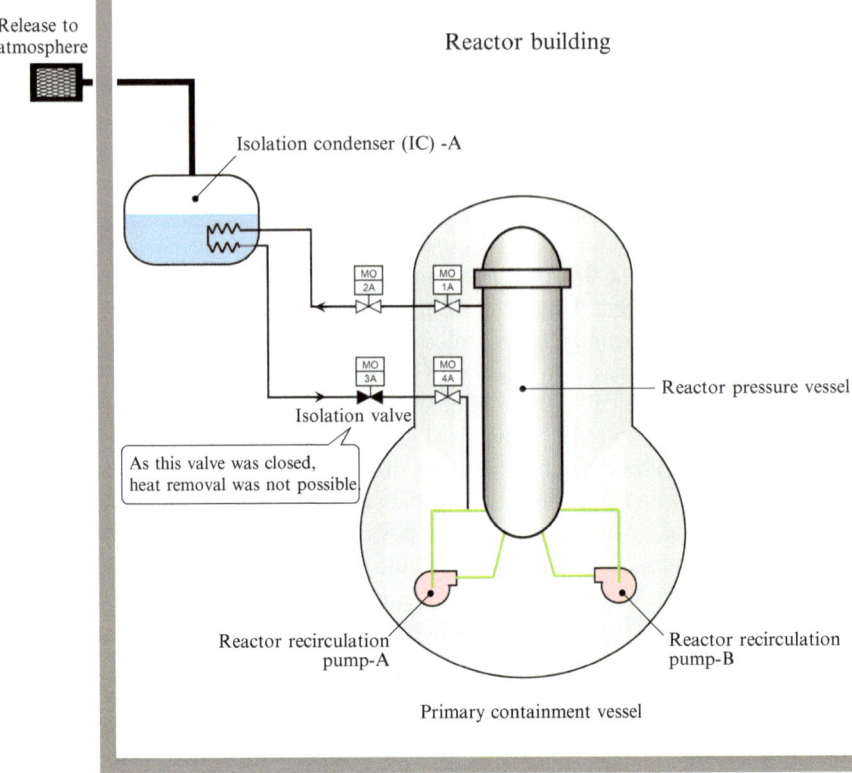

Fig. 2.5 System diagram of Unit 1 isolation condenser

device using gravity so that if this IC worked as we expected, the core melt and hydrogen explosion of Unit 1 would not have occurred and the accident would have been much smaller.

It is so designed that as much as 100 tons of water is stored on the secondary side of IC's heat exchanger, so that it can continue to cool the reactor for 8 h continuously when the reactor stops. Of course, the cooling period can be easily extended by simply replenishing the water. Unit 1 was equipped with two ICs, i.e., A and B.

Although I will be explaining later why the ICs did not work as expected, the problem is that neither the Site Response Headquarters nor the Head Office Response Headquarters of TEPCO were aware of the fact that the ICs were not working and simply took it for granted that the ICs were working. I suppose that this misunderstanding amplified the accident. It seems that the main concern of TEPCO headquarters immediately after the tsunami hit was focused on Unit 2, whose output power is larger and whose reactor water level and operating status of RCIC were unknown at the time, and they paid little attention to Unit 1, thinking that it was safe, as its output is smaller and has natural circulation cooling ICs.

The reason that these ICs of supposedly higher reliability did not work is that the isolation valve (MO-3A), which is located where the piping that connects the reactor with ICs pass through the containment vessel, was closed.

There are many valves related to ICs. Each valve's status during the accident, i.e., whether the valve worked or not, is still in the midst of discussion. Since it deviates too far from the purpose of this book to discuss each of them in detail, I will focus my discussion of the performance of ICs focused on the status of MO-3A valves which the operator controlled.

Why then was this MO-3A valve closed?

The reactor was automatically shut down as the earthquake hit the power station at 2:46 p.m., March 11. With it, the power outage occurred and the main steam isolation valve was closed automatically. However, decay heat continued to be generated even though the reactor was shut down. As the main steam isolation valve was closed, the steam generated by decay heat had no place to go so that it ended up causing the reactor pressure to rise. At 2:52 p.m., 6 min after the reactor's shut down, the ICs automatically kicked in as designed, having received the reactor pressure high signal. Unit 1 then switched to the cooling by ICs.

The cooling speed is typically very fast immediately after ICs are kicked in. This is because the cool water that has been kept in the cooling tube of ICs rush into the reactor all of a sudden. I personally experienced it during my days at JAERI witnessing how IC of JPDR works; it really cools the reactor well immediately after it is kicked in. I heard that they used to conduct emergency trainings at Tsuruga NPS using IC from time to time during the cooling process after scheduled reactor shutdowns, but it looked that this was the first time for the operators of Fukushima Daiichi Unit 1 to experience this.

At 3:03 p.m., 10 min after the IC operation started, they suspended the IC cooling because they thought that the cooling speed was too high. They used the particular MO-3A valve to stop it. This was not a wrong operation. The operation rule states that the cooling speed of the reactor shall not exceed 55 °C/h. At the particular instance, the operating staff had no idea as to whether a tsunami would attack them later so that they simply obeyed the rule. It was the standard procedure to do so.

Not much later, the temperature and pressure of the reactor began to return to normal, the operating staff kept only one of the IC systems operating, and operated the isolation valve (MO-3A) on the condenser return piping on and off in order to control the cooling speed of the reactor manually. It seems that this operation was done several times. It was quite unfortunate that the tsunami came immediately after the MO-3A isolation valve was closed by the operating staff. The electric power of Unit 1 was totally lost by tsunami. The electric power to operate MO-3A was gone. Thus, the ICs became inoperable.

As I mentioned above, there have been many things said about the status of IC of Unit 1 after the accident. I have my own thought about it, but that is not the purpose of the book, i.e., the analysis of the core melt, so that I do not wish to discuss about it here.

However, it is still true that whether the ICs worked or not was detrimental in the occurrence of the core melt. It was a critical issue of grave consequence that determined the life and death of the reactor. The operating staff must have known that. However, what attacked them was earthquake, tsunami and total blackout due to power failure. It would be too harsh to demand perfect performance for the operating staff on all matters such as emergency operating performance and accurate information recording and transmission. If only they had a portable generator or a battery, MO-3A could have been opened. If they had such preparation, the operating staff could have been more attentive. I will be discussing this in Chap. 6 of Part II.

The stoppage of ICs made a night-and-day difference in the Fukushima Daiichi accident. Their late recognition of the IC stop was the biggest mistake they made. However, crying over spilt milk does not move us forward. Let us move on now, admitting that the core cooling of Unit 1 is lost forever as the IC function is lost by the attack of tsunami.

2.4.2 Fuel Temperature Rise

The second wave that flooded the power station arrived at 3:35 p.m. The ICs became inoperable about 50 min after the earthquake.

However, it was very lucky for Unit 1 that the ICs operated for at least this period of 50 min. It is because the decay heat which used to be about 7 % of the rated power immediately after the reactor shut down was reduced to about 2 % in 50 min. In other words, the size of the decay heat that triggers the core melt was reduced to 2 % at the starting point. It means that a large advantage was given to the operating staff at the start, so that the time to the core melt was substantially elongated in comparison with the case of TMI.

Unfortunately, TEPCO failed to take advantage of this luck. It is because they did not realize that the critical device, the ICs, were not operating. If they were aware of it, there were ways to take advantage of it. Many things could have been done since the radiation level of the site was low enough so long as they could manage the darkness.

Since IC is a simple natural circulation circuit, the reactor can be cooled even without electricity, so long as the valve is open. That's why the commander and his staff at the site headquarters did not pay too much attention to Unit 1 assuming that the ICs were operating. That was their critical mistake.

I assume that it was around 10–11 p.m. of March 11 when they noticed that the radiation level inside the reactor building rose sharply when Unit 1 was in a critical condition.

According to TEPCO's analysis, the reactor's water was empty about that time. However, this conclusion can be discounted slightly. It is because the γ-ray that is emitted from the fuel rods do not function as the source of the decay heat as it disperses to the outside of the core when the reactor water level drops and the fuel rods become exposed above the water level.

The discussion of this reason takes a lot of words and deviates from the main subject so I will not discuss it here, but it contains some important matters in considering the nuclear reactor accident so that it will be written as at the end of this chapter as Appendix 2.1. You are welcome to look at it if you are interested in.

Now, let us summarize what transpired with respect to Unit 1's core after the ICs ceased to operate.

The ICs stopped about 1 h after the reactor's shut down. The magnitude of the decay heat is approximately 2 % of the rated power, or more specifically, approximately 30 MW. With this thermal energy, the water in the reactor core is heated to produce steam, resulting in an increase of the reactor pressure. When the pressure rises about 10 %, the safety relief valve opens automatically to discharge the steam to the containment vessel, and restores the reactor pressure back to about 7 MPa. When the pressure returns to normal, the valve closes. This steam blowout-and-stop cycle repeats itself again and again after the isolation condensers ceased to function.

This steam blowout by the safety relief valve is essentially the cooling of the reactor by evaporating the water kept inside the reactor. It can be compared to an octopus eating its own arms in order to survive. It cannot last long. It consumes the water in the reactor and keeps lowering its water level. On the other side, the temperature and pressure of the sump water in the containment vessel where the steam is dumped increase as the amount of the dumped steam increases.

Let's make a simple calculation. The water evaporation by this heat is calculated as approximately 75 tons/h. According to TEPCO's calculation, the water level that used to be approximately 5 m above the core dropped to the upper portion of the core 3 h later (about 6 p.m., March 11), and reached the bottom of the core about 5 h later (about 8 p.m.) (Fig. 2.6). Since this is a simple calculation of the water level drop due to steam generation, it is slightly exaggerated as I mentioned before, but it can be trusted as a ballpark calculation. As the time required from the water level to drop the length of the core (approximately 4 m) was approximately 1.5 h, the water level drop speed at the core was approximately 2.7 m/h, or 4.5 cm/min.

Now let us look at the fuel status when the water level has dropped to the middle of the core, in order to compare it with the TMI accident.

The water inside the reactor was only boiling calmly during the period when the safety relief valve was closed. The fuel rods at this moment were like bathers in a sauna bath just like the fuel rods of the TMI accident when the block valve of the pressurizer relief valve was closed. Since the evaporation speed of steam is roughly 1.5 cm/s or so (Appendix 2.2), it was a sauna without any breeze. However, when the pressure increased and the safety relief valve was open, the steam vented from the reactor so that the fuel rods were cooled somewhat. While the portion of the fuel rods that was exposed above the water level was sometimes heated by the sauna and sometimes cooled by the steam flow, its temperature must have gradually risen as the water level dropped further.

It must be that the fuel cladding tubes exposed above the water level began to be oxidized and covered by a thin oxide film on the surface. I assume that the inside of the softened cladding tubes must have adhered to the pellets and formed oxide film as well. I believe it is safe to assume that it was no different from the status of the TMI accident but the fuel status of Unit 1, whose decay heat was less than that in

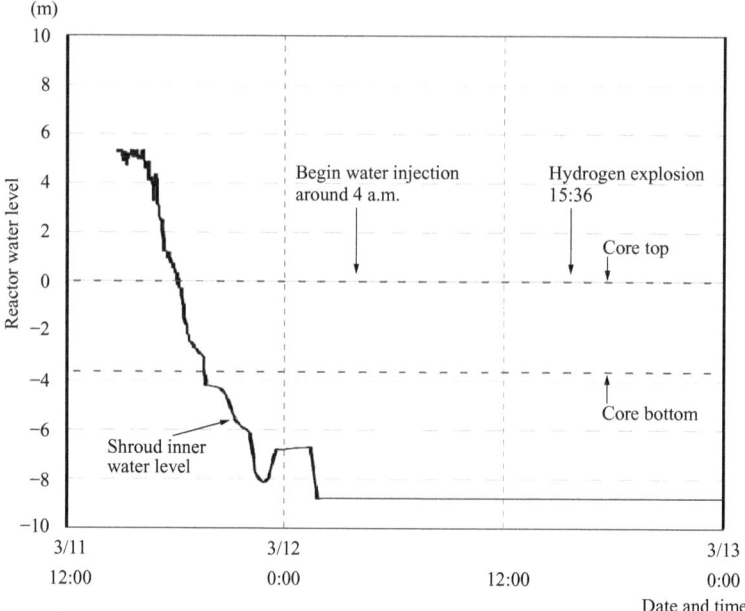

Fig. 2.6 Unit 1 water level change (analysis data) (Source: from TEPCO "Fukushima Nuclear Accident Investigation Report")

the case of TMI, was in a status between [II] and [III] of the model diagram (Fig. 1.5) of Sect. 1.4.

In case of TMI, the status of the fuel rods was at [IV] as the decay heat was high. That was when the reactor coolant pump began to run. Hence, the core collapsed. Please refer to Sect. 1.5. A large amount of cool coolant water flowed into the core, the fuel rods broke up into pieces, the core collapsed, and the core melt occurred. However, in case of the Fukushima accident, it was about 7 p.m. when the water level dropped to a half, but the pump did not run due to the power outage, so that water did not flow into the core. Therefore, neither fuel rod breakdown nor core collapse occurred. The fuel rods tenaciously held the status of either [II] or [III] of Fig. 1.5.

This condition continues. However, the further the water level of the core dropped, the length of the portions of fuel rods below the water became shorter, so that the water evaporation amount reduced, and the degree of overheating in the upper portion of the core increased proportionately. In other words, the amount of heat used for evaporation declined, and while that to raise the steam temperature increased. Consequently, the fuel temperature in the upper portion of the core rose by that much. The status of the fuel rods probably shifted gradually from [III] to [IV].

By midnight of March 11, the water level of the reactor was completely lost. Although the decay heat generated in the core needs to be subtracted somewhat, due to the increased heat dissipation by radiation with the temperature of the fuel rods.

From this point on, the situation requires us a comprehensive study on the status of the fuel rods the temperature of which continues to rise while they keep emitting radiation heat, of the temperature increases of the internal components of the core and the surrounding structures such as the core structure, the reactor pressure vessel,and the containment vessel exposed to the radiation. The temperature evolution of each component is determined by the give and take balance of the radiation heat, in addition to the core melt.

As the water level drops to a half of what it used to be, the TMI accident is no longer useful as an example to learn from. From this point on, it is a voyage without a nautical chart, and there is no way out but to think carefully about the condition of the core on our own in order to find what happened. As I laid out, there are so many things to think about.

Incidentally, the most complex case of core melt among those of Units 1–3 is that of Unit 1. Therefore, I wish to temporarily stop the analysis of Unit 1 here, and do the analysis of the core melts of Unit 2 and Unit 3 which are closer to the example, i.e., TMI, first and then resume the Unit 1 analysis based on those examples. I believe the explanation will be easier for me and it will be easier for the readers to agree with my thought as well.

2.5 Case of Unit 2

Unit 2 was completed and commenced operation in 1974, 3 years after Unit 1. The main difference between the two units is that the output is 460 MW in Unit 1, and 784 MW, or approximately 70 % larger, in Unit 2. As the output increased, so did the physical size of each facility, but the design concept itself did not change much.

If we are forced to find major differences, the differences are, as I mentioned for Unit 1, that the reactor core cooling system with respect to isolation cooling (cooling when the reactor is isolated from the main turbine and condensers) was changed from IC to RCIC, and that many of the control systems were converted to electric type. Another marked difference is that most of the equipment, which were made by GE in case of Unit 1, were switched to Japanese makes such as those made by Hitachi and Toshiba. However, you should understand that there is no difference between Unit 1 and Unit 2 as to the core structure, which is the key subject of this book.

2.5.1 Reactor Core Isolation Cooling System (RCIC)

Although Unit 2 was affected by the power loss and the reactor was shut down as a result of the earthquake, the cooling process was started by the operating staff using the Reactor Core Isolation Cooling (RCIC). RCIC is a cooling system used when the reactor is isolated as mentioned above, same as the ICs used in Unit 1.

The difference between RCIC and IC is that, while IC uses natural circulation for cooling, RCIC uses a steam turbine driven pump for cooling, so that it needs to be controlled electrically.

The RCIC is capable of maintaining its own mechanical power because the steam used for driving the pump is generated by the decay heat of the core and does not depend on electric power. While the RCIC can control the water level, it requires DC (battery) power for the control. The balance between merits and demerits is a delicate one. Unfortunately, the scale tipped in the wrong direction in the accident. The battery was flooded and the system became uncontrollable.

Let me explain how RCIC works.

In case of a normal reactor shutdown, the steam generated by the decay heat is discharged to the turbine condenser to be cooled by a large amount of seawater. However, in case of this accident, the circulation water pump for pumping up seawater was disabled not only by the power outage but also by the damage caused by the tsunami, so that there was no way to discharge the steam to the turbine condenser. That's where RCIC comes in.

The RCIC pump has two water sources: the suppression chamber ("SC"), which is a very large water sump located in the lower part of the containment vessel, and the condensate storage tank. However, as the water of the condensate storage tank was never used in this accident since the switching was made on the early morning of March 12, only the SC will be considered as a water source in the following discussion. Let's look at Fig. 2.7.

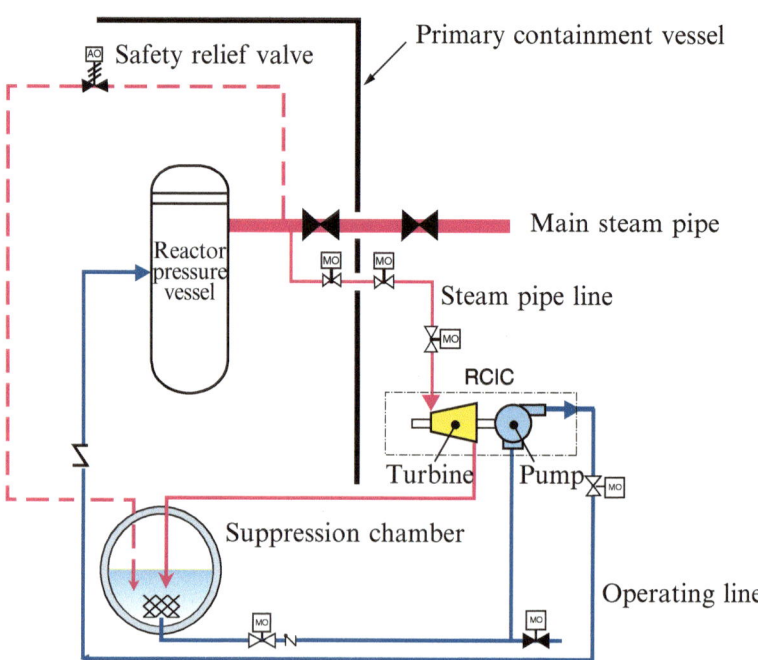

Fig. 2.7 Reactor Core Isolation Cooling (RCIC) system diagram

The water from the pump enters the reactor pressure vessel via the water supply piping and replenishes the water lost by the decay heat. A battery is needed to control this supply amount to maintain the water level of the reactor properly. On the other hand, the steam generated by the decay heat enters the SC again after having propelled the turbine that drives the RCIC pump and having been condensed back to water.

In other words, the reactor cooling by RCIC forms a closed loop circuit by using the water of the SC, i.e., the sump of the containment vessel, as the cooling water, while returning the steam generated in the reactor back to the same SC. In other words, the reactor cooling using RCIC is [a circulation cooling] by means of the SC water of the containment vessel, and so there is a net zero effect on the water balance. That means, however, that the cooling capacity (capability) is governed by the quantity of the SC water. The SC water amount that can be stored in the Mark-I type containment vessel is approximately 3,000 tons. If we assume that RCIC can be used until half of that amount of water is evaporated, it will be able to cool the reactor for 5 days or so.

By the way, most of the reactors of Fukushima Daini NPS and Onagawa NPS used this RCIC for the core cooling after their reactors were shut down during the Great East Japan Earthquake. Although the operating staff had a very tough time due to the damage caused by tsunami, they were eventually able to achieve cold shutdown status on all of those reactors.

As can be seen from this example, reactors can be cooled so long as water injection from RCIC can be continued. Since the decay heat is only about 7 % of the rated output, the temperature of the fuel rods do not rise. The temperature of the fuel rods is about the same as that of the cooling water, i.e., 300 °C or so, so that it is far removed from the world of fuel melting and oxide films forming on the cladding tube. It is safe for us to think that a core melt does not occur so long as the RCIC is functional ((a) of Fig. 2.10).

At around 3:35 p.m., Unit 2 was also hit by the tsunami. The tsunami flooded the battery and made it impossible to control the RCIC. However, all of the main units of RCIC were installed on the reactor building which was saved from the tsunami, and it was lucky that the turbine that drives the pump continued to run without the help of the control power and kept supplying water to the reactor.

A TEPCO report states that it was confirmed that the reactor water level was 3,400 mm above the core as of about 10 p.m. on March 11, and that the operating staff confirmed that the RCIC operating in the blacked-out site at around 3 a.m. on March 12. That must have been a heartening report for TEPCO. The people at the site may have hailed, "OK, we can make it!"

What should be noted is that the water level of plus 3,400 mm that was confirmed on the night of March 11 was kept unchanged until around 10 a.m. of March 14, as evidenced on the TEPCO's water gauge record (Fig. 2.8). This means that the RCIC pump kept running without control for 3 days since the accident. This is a total surprise, because an isolation cooling facility like RCIC is normally designed for a power outage of 8 h. A machine designed to run 8 h kept running for three complete days in a poor environment without control. We are badmouthed because

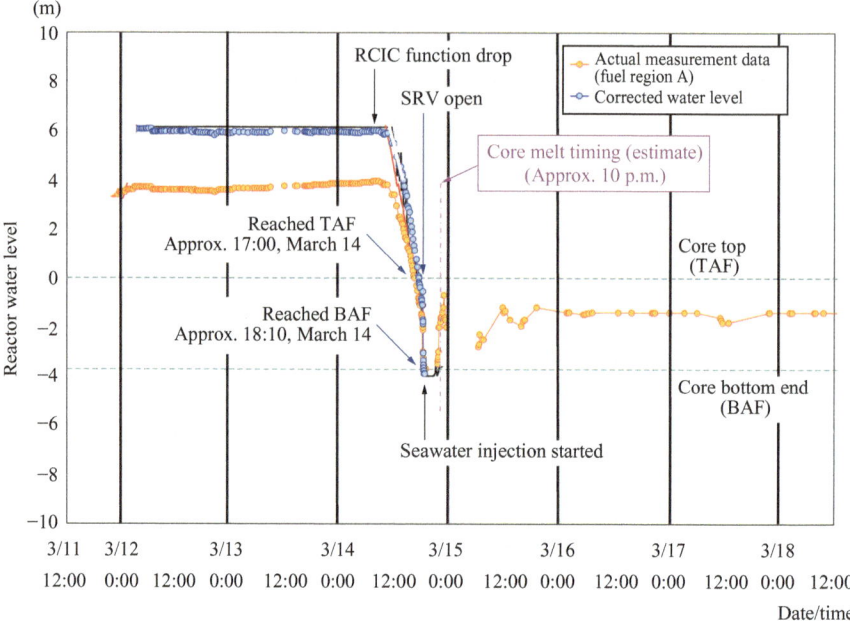

Fig. 2.8 Unit 2 reactor water level change (measured data) (Source: from TEPCO "Fukushima Nuclear Accident Investigation Report")

of the accident, but our safety devices at the power plant stood up for their names. They worked better than they were expected. Incidentally, this 8 h design criteria is common to all nations around the world, except for a few exceptions.

According to the report, this plus 3,400 mm water level amounts to plus 6,000 mm if the error of the water gauge due to the accident is compensated for. Although I will not elaborate on the reasoning, I trust this compensation is correct. If the water level was actually plus 6 m, it exceeds the top of the steam-water separator, so that it is reasonable to expect that the quality of the separated steam was poor and contained water. I suppose that the rotary blades of the RCIC drive turbine were substantially damaged by the water content of the steam.

The reactor water level started to drop after 11 a.m. of March 14. The report states that the power plant manager judged that the RCIC had ceased to function, noting the extreme drop of the water level, at 1:25 p.m. on March 14. I suppose that the RCIC pump stopped at around 11 a.m. (Appendix 2.3).

From that point on, the decay heat of the reactor was removed by water evaporation, and the reactor's water level started to drop because of it. In other words, the core cooling condition of Unit 2 became the same as that of Unit 1. It is safe to say that the core melt problem of Unit 2 started around 11 a.m. on March 14.

Coincidentally, an explosion occurred in the reactor building of Unit 3 at 11:01 a.m. on March 14. Because of this explosion, the temporarily installed water hose and fire engines prepared for the purpose of feeding water to Unit 2 were damaged and became unusable. In addition, the operating device of the vent valve prepared for pressure relief of the containment vessel became unusable as well, so that the pressure venting of Unit 2 became difficult. As the on-site task force team of Unit 2 was busy coping with these matters, the water injection to the core was delayed until 8 p.m. of the same day.

This delay caused the core melt, which I will be discussing later.

Let's get back to the discussion of the core. When the RCIC pump stops, the water injection to the core stops, so that the status of the fuel rods become a cooling state as if they are bathing in a sauna bath. This causes the core water to evaporate, the fuel rod temperature to rise, and the oxidation of the cladding tubes to start in no time.

As we look at Fig. 2.8, the water level, which used to be plus 6 m at 10 a.m., March 14, reached exactly to the bottom of the core, i.e., minus 4 m by 6 p.m. of the same day. The cooling water is totally gone from the core by around 6 p.m.

Let's make sure if this understanding is correct by checking the data from 11 a.m., when the RCIC stopped, to around 6 p.m., when all the water of the core is gone. As far as we can see from Fig. 2.8, the core water level decreased very smoothly without any unexpected change. The reactor pressure (Fig. 2.9) started to rise at around 10 a.m. because of water evaporation, and was kept at around 7.5 MPa constantly by means of the operation of the safety relief valve from a little after 2 p.m. until approximately 6 p.m. All the data look as expected with no special deviation.

The fact that no irregularity can be found in the reactor's pressure data means that the core was cooled to a certain degree until 6 p.m., and no problem such as a core melt occurred. We assume so because, if core melting occurred, there must

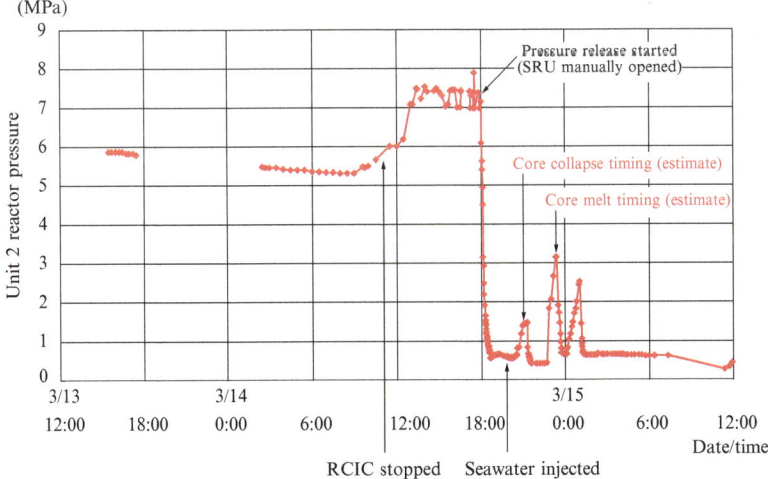

Fig. 2.9 Unit 2 reactor pressure change (measured data) (Source: from TEPCO "Fukushima Nuclear Accident Investigation Report")

have been an acute core water level change due to the enormous heat caused by the zirconium-water reaction. It also must have caused an abrupt reactor pressure rise by the generation of hydrogen. The fact that there was no abrupt change as such proves that there was no core melt until 6 p.m.

Although this may sound very obvious, it has a very important meaning. It means that core melt did not occur even though there was no water in the core and it was exactly like heating a bath with no water in it. This is a fact that was clarified for the first time in this Fukushima accident, and it was a new finding that was not experienced in the TMI accident. This, of course, is data that should be utilized in reactor safety from now on.

Now, we know that the water level was down to the lowest part of the core by 6 p.m. I have a view that the fuel temperature increase due to the water level decreased was not so high, probably reaching only up to 1,000 °C at most, because the decay heat had reduced to about 0.4 %, different from the case of the TMI accident, and also because there was a heat dissipation of breezy intermittent steam flows due to the safety relief valve. The state of the fuel rods was probably the [I] or [II] stage of Fig. 1.5 at most.

The core has not started to melt by this time, i.e., the early evening of March 14. The core melt has yet to start at this point.

2.5.2 Sea Water Injection and Core Melt

At 6:02 p.m., the safety relief valve was fixed to an open position to lower the reactor pressure.

At 7:54, the water injection to the reactor was started using fire engines.

The combination of these two operations triggered the core melt.

The title of this section is the core melt. In order for the core melt to occur, two conditions must hold simultaneously, i.e., that the fuel rods are at a red-hot state and that a large amount of water is available, as described in the conclusions I and II of Sect. 1.7. Please keep these in your mind as you read the following.

Before we get into the main discussion, let us make some calculations that are needed for assessing the situation.

As I described in Sect. 1.4, the size of the decay heat that was a partial cause of the core melt within 2 h after the shutdown in case of the TMI accident was approximately 1 %, and the rate of temperature rise of the fuel rods due to this heat was a little over 0.74 °C/s. In case of Unit 2, the decay heat had dropped further to about 0.4 % as it had been a full 3 days after the shutdown. If we assume that no water was in the core, so that no heat was being removed at the time, the rate of temperature rise of the fuel rods was about one third of that of TMI, or 0.2 °C/s, or 12 °C/min, or approximately 700 °C/h.

At 6:02 p.m., the operating staff lowered the core pressure forcibly by firmly setting the safety relief valve to the open position. The reactor pressure had dropped sharply from about 7.5 MPa to about 0.5 MPa. The time needed for the pressure drop was probably 30 min or so.

Due to this pressure drop, the water in the lower part of the reactor began to "decompression boil." Decompression boiling is a kind of self-boiling phenomenon caused by a decrease of the saturation temperature associated with a drop in pressure. The consumption of cooling water by this decompression boiling can be substantial. According to my ballpark calculation, the decompression boiling caused approximately 30 tons of water that was below the core to evaporate. Those of you who are interested may review the calculation provided at the end of this chapter (Appendix 2.4).

Thus, the water level dropped to approximately 1 m below the core.

In addition, it is assumed that the core fuel that had been kept at around 1,000 °C in a sauna-like condition up to that point was cooled by the steam generated by decompression boiling and the temperature dropped to around 150–160 °C, or close to the saturation temperature of water.

In other words, the fuel rod temperature dropped because of the decompression boiling. If seawater was injected into the reactor immediately after the completion of the pressure reduction at this time, the core melt would not have occurred. That is because zircaloy covered by oxide film does not react with water if the temperature is low. If the zirconium-water reaction does not occur, a core melt will not occur even if the fuel rod splits up and fall down, because such fuel debris will be cooled by water ((c) of Fig. 2.10).

Please recall the PBF-PCM test described in Sect. 1.3, the communication path, and the agglomeration of cooled fuel debris found on top of the TMI core.

Unfortunately, it was as late as 7:54 p.m. when seawater finally arrived at the core. It was approximately 2 h after the pressure reduction started. It can be easily estimated from the ballpark calculation we did a while ago that the temperatures of most of the fuel rods must have reached around 1,500 °C by then.

The surface of the fuel rods sticking out in a red-hot state above the water surface must have been covered by oxide films, and the inside of the oxide films must have been filled with zircaloy, i.e., the cladding tube material, which was softened, partially melted, and formed agglomerations here and there. Then came a flush of cold water. If the seawater was sprayed from the top of the core, the oxide film would have been quenched and broken down, causing the fuel rods to split into pieces, so that the core must have collapsed judging from the precedents of PBF and TMI. The core could have collapsed instantly into a pile of debris. But it didn't turn out that way.

The water was not sprayed from the top of the core, but was fed from the piping on the side wall of the reactor pressure vessel. Besides, it was a tiny flow from a fire engine. A portion of the seawater that streamed downward along the wall may have evaporated but the majority of it must have accumulated into the bottom part of the reactor pressure vessel serving to raise the rector water level.

Checking with a fire station, I found out that a fire engine pump takes 3–5 min to discharge the water contained in its 2 ton tank. Since this amounts to 24–40 tons/h, it takes about an hour to replenish the water lost by decompression boiling.

Please look at the reactor pressure change diagram of Fig. 2.9. A small but sharp pressure increase can be observed at around 9 p.m. of March 14. Around 9 p.m. corresponds to about 1 h after time the water injection started. This is about the time

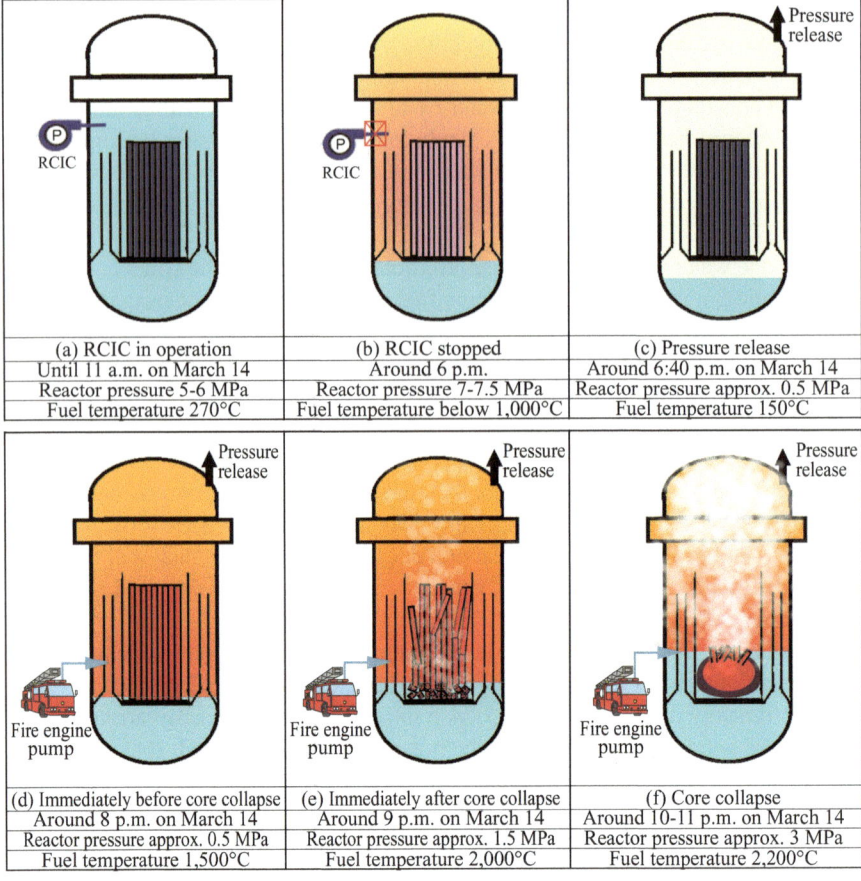

Fig. 2.10 Progress of Unit 2 core status (model diagram)

the reactor pressure vessel's water level is estimated to have reached the core bottom. I assume that it is about the time when a certain amount of zirconium-water reaction occurred as the water level reached the bottom of the hot fuel rods ((e) of Fig. 2.10).

I imagine that the water that contacted zirconium partially boiled, splashed around, and evaporated. Hot fuel rods, on the other hand, split into pieces as they contacted cold water, and the molten zircaloy reacted with water to generate hydrogen. I imagine that this generation of hydrogen caused a temporary reactor pressure increase of about 1.5 MPa.

However, the reaction was not so large as to cause the core to melt. The reason was that the amount of water available for the reaction was insufficient.

In the TMI accident, it took only about 2 min for the core to melt in the TMI accident. Although the primary coolant pump was operated for only 19 min, it is thought that approximately 28 tons of coolant was injected into the core within less than a minute because a large amount of coolant stored in the piping was injected by a large capacity pump. With this large amount of water gushing in, it is believed that the core was, thus causing the top portion of the core to break into pieces and ending up in an overall collapse of the core, as well as causing the zircaloy left in the core to oxidize altogether to generate a large amount of hydrogen.

On the other hand, in the case of Unit 2, water was supplied in a small stream from a fire engine. Since the pumping speed from a fire engine is typically 24–40 tons/h, it was only one sixtieth of the rate of water supply in the case of the TMI accident. If not enough water is supplied, an oxidation reaction cannot occur. Thus, the core melt did not occur in a short period of time as in the case of TMI. A further supply of seawater was needed in order to cause the core melt to occur.

Please keep in mind that what is meant by a core melt here is a first violent reaction between water and the mixed molten substance in the core and amounts to about 30–40 % of the total zircaloy. The reaction between water and the remaining zircaloy is assumed to have occurred later from time to time depending on the circumstance.

Let us do some brush-up now. We have learned that the heat that causes the core to melt is not decay heat but rather an enormous amount of reaction heat generated by the oxidation reaction between water and the zirconium used as the cladding tube material. However, in order to cause this reaction to occur all at once, both a high fuel rod temperature and as well as a large amount of water are needed. The amount of water supplied by the fire engines was not sufficient to cause the reaction to occur all at once.

Let us make a calculation from a different angle. The total amount of hydrogen generated by Unit 2 is estimated as 460 kg by TEPCO. Let us use 500 kg as the approximate number in our calculation. Since the molecular weight of hydrogen is 2 and the same for water is 18, it takes 9 water to generate 1 hydrogen in the weight ratio. In order to generate 500 kg of hydrogen, it requires at least 9 times of it, i.e., 4.5 tons of water. As I am not sure how much of the total injected water contributes to the reaction, let me be conservative and assume that 50 % of it contributes to the reaction; then the required water would be approximately 9 tons. In reality, the required quantity is probably more than that. That much amount of water requires at least 15–30 min to be supplied from a fire engine.

The above is only a rough estimate but is enough for you to realize that it takes a substantial length of time for Unit 2 to develop a core melt, different from the case of the TMI accident where it was caused by a large amount of water supplied in one quick shot.

Actually, the evidence is in the reactor pressure diagram of Fig. 2.9. The reactor pressure curve made sharp temporary rises three times between 9 p.m. of March 14 through the early morning of March 15. Although I cannot be too certain because the time scale is rough, I can only assume that these irregular pressure rises are caused by violent hydrogen gas generation due to zirconium-water reactions. I also assume that the core melt of Unit 2 occurred three times intermittently between the evening of March 14th through the morning of the 15th.

The time needed for the first peak to occur is the time required for replenishing approximately 30 tons of water, the amount of water lost by the decompression boiling. As can be seen from Fig. 2.9, the fact that the first pressure rise occurred approximately 1 h after the water injection is convincing proof.

Let us estimate the core condition based on these pieces of circumstantial evidence. The fuel rods cooled down for a while due to the steam discharge caused by the forced decompression that started at 6:02 p.m. on March 14, but it took a total of 3 h for the seawater to be delivered to the bottom of the core – a 2 h delay in the delivery and another hour to be delivered to the destination. During this period of time, the fuel rod temperature rose to approximately 2,000 °C.

The fuel rods still stood straight up like forest trees, however. Then comes in the seawater pumped in by the fire engine. As the cold seawater from the fire engine started to flood the lower part of the core, a violent boiling occurred and the sloshing seawater, violently sloshing because of the boiling, quenched the surface of the fuel rods. I am sure that this caused the oxide film of that part to break up, causing the fuel rods to disintegrate and collapse. In other words, a partial core collapse occurred.

In the meanwhile, the debris of the broken-up fuel rods was piled up on the core plate. There is no question that some of the agglomerations of high temperature zirconium, which were protected by the oxide films formed here and there, streamed out through cracks formed by the breakup and caused oxidation reactions as they met with water. There is no question that, as a result of this heating, localized melting occurred in the lower part of the core and generated hydrogen as well. This hydrogen is the cause of the first rise in pressure.

I suppose that the fuel rods that collapsed because of the loss of support still remain there, holding their shapes from the middle to the top of the core. There must have been an egg-shell made of the mixed molten substance in the vicinity of the core bottom to protect the molten core from making direct contact with the water.

This area is where we can make various speculations and conjectures. It is a world of chaos. I think you should use your imagination freely to picture it. It can be seen from Fig. 2.9 that such a chaotic world lasted for a few hours.

I suspect that, after a couple of hours of such a chaotic condition, the fuel rod temperature's rising, and the water level of the lower part of the core rising again to a sufficient height, the direct contact between water and the fuel rods restarted. This is when the full-scale core melt started. Sharp reactor pressure rises that reached 2.5–3 MPa are recorded twice between the late evening of March 14th through the early morning of the 15th. There is no question that large zirconium-water reactions occurred.

The second reaction was around 10 p.m. The radiation dose rate in the vicinity of the front gate jumped up sharply just about this time. I estimate the time of the core melt of Unit 2 to be the midnight of March 14, although I cannot determine the exact time ((f) of Fig. 2.10).

2.5.3   *Hydrogen Gas Generation and Radioactive Release*

Now, let us look at the condition up to this point from the perspective of the pressure change of the containment vessel (Fig. 2.11). The containment vessel comprises a space called a dry well ("DW") that surrounds the reactor vessel and a doughnut shaped part called a suppression chamber ("SC"). The DW and SC are connected by an interconnecting pipe. The SC contains a large amount of water and there is a vacant space above the water. The steam that is discharged from the safety relief valve is sent directly to the SC and is cooled by the SC water to return to water.

Next, let us think about hydrogen gas discharged from the reactor. The hydrogen gas discharged from the DW passes through the interconnecting pipe, the SC water, and enters into the SC vacant space. The vacant space has a normally closed vent piping. In the middle of the vent piping provided is the rupture disk in question. In case of an emergency, the gas that enters the SC can be guided to rupture the rupture disk to escape to the atmosphere via the stack by opening the vent piping.

That is the explanation of the Mark-I type containment vessel. Although the DW is also equipped with vent piping, it will not be discussed here because it has nothing to do with the accident in question.

Incidentally, each of the DW and the SC is equipped with a pressure gauge. The two gauges normally indicate about the same pressure. Since both of them indicated same pressures until midday of March 14, we have no problem in trusting them (Fig. 2.11).

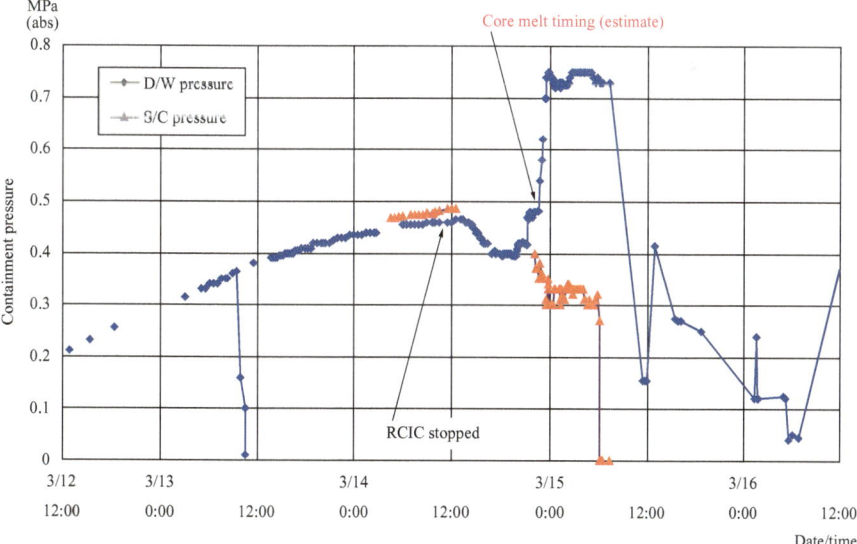

Fig. 2.11 Unit 2 containment vessel pressure change (measured data) (Source: from TEPCO "Fukushima Nuclear Accident Investigation Report")

However, only the data of the DW was available since then until after 10 p.m. of the same day. To our chagrin, when the SC pressure was restored at around 10 p.m. that day, its indication was entirely opposite to the DW pressure and became unmeasurable from around 6 a.m. the following day. This is a problem. Both of these data are not perfect for explanation, but I will be using the DW data as the correct containment vessel pressure data in the following explanation, but I suspect that both gauges were damaged by sharp pressure rises.

The reactor pressure (Fig. 2.9) gradually started to rise approximately 2 h before the RCIC pump stopped, i.e., about 9 a.m., and the containment vessel pressure (Fig. 2.11) started to drop around 1 p.m. This shows the fact that the RCIC pump began to run abnormally, and the steam quantity consumed by the pump (for reactor heat removal) began to drop. As a consequence, the reactor pressure tends to rise, while the containment vessel pressure tends to drop due to the reduction of steam quantity entering the SC water.

At around 2 p.m., the reactor pressure rose to approximately 7.5 MPa and steam began to blow out intermittently from the safety relief valve. The DW pressure tends to decrease slightly at this point. This is believed to be caused by the reduction of the mass flow, because the gaseous phase flow became dominant with the drop of the water level. When the water level was high, the liquid flow had been dominant, even if there was no change in the reactor pressure. We can see that the containment vessel was discharging a large amount of heat from the fact that the containment vessel's pressure was reducing despite the fact that mass and energy were flowing in (Fig. 2.12).

The DW pressure indication slightly upward starting around 8 p.m., but at 10 p.m., it began to turn sharply upward from 0.4 to 0.8 MPa. This sharp rise represents the harshness of the hydrogen generation due to the core melt. In case of Unit 2, the DW volume of the containment vessel is approximately 4,000 m^3. In order to raise the pressure of this volume by 0.4 MPa in one stroke, a humongous amount of gas is required – 16,000 m^3 of gas, if converted into atmospheric pressure.

Since hydrogen gas is generated from the molten core, the temperature is about same to that of the molten core, i.e., 2,000 °C. If a volume of 16,000 m^3 is to be filled with hydrogen gas of this temperature, approximately 200 kg of it is necessary (Appendix 2.5). That amounts to approximately 40 % of the total amount of hydrogen generated by Unit 2 as discussed in Sect. 2.5.2. This sharp pressure increase can only be explained by the hydrogen gas generation. To put it in another way, the existence of this pressure rise is the evidence of the core melt time.

By the way, the design pressure of the containment vessel is approximately 0.4 MPa. However, it is confirmed by various tests that it can withstand pressures about twice that pressure from the standpoint of structural strength. Having said that, however, I am impressed that it withstood the high temperature hydrogen gas of approximately 0.8 MPa.

Now, the question is the temperature of the containment vessel. There is no such data. TEPCO has calculated the containment vessel temperature at this time to be 150–170 °C. The hydrogen gas release occurred from around 9 p.m., after having

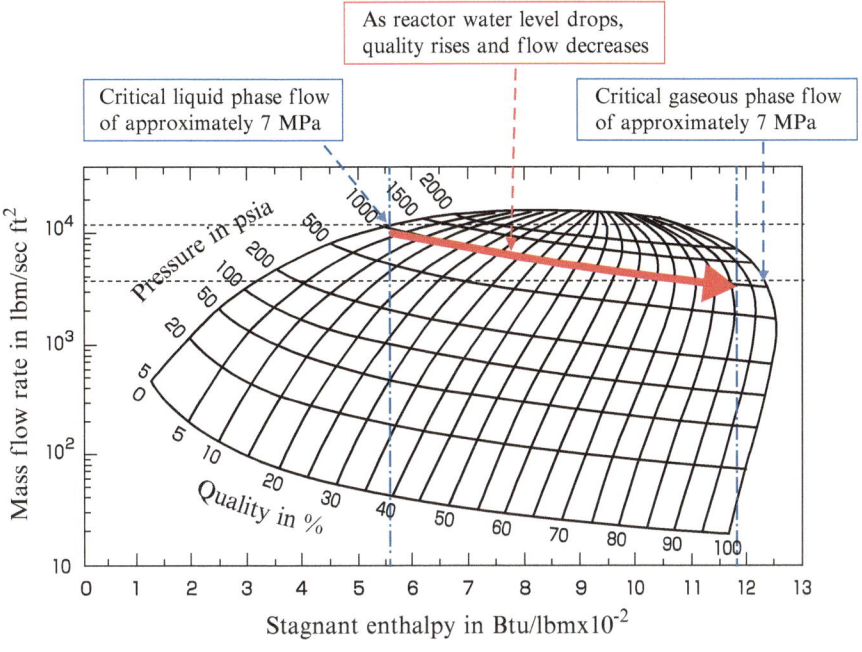

Estimated critical two phase flow rate

Fig. 2.12 Mass velocity and quality (Source: Fauske, H. K., "Two-Phase Critical Flow," Paper presented at the M. I. T. Two-Phase Gas-liquid Flow Special Summer Program, 1964)

this preheating. I estimate that the temperature of the upper portion of the containment vessel must have been at least the reactor saturated steam temperature, i.e., 300 °C or even higher.

With such a high temperature, what is concerned is thermal expansion. If the bolts that are used to fasten the lid of the containment vessel elongate due to thermal expansion, the fastening of the lid gets loosened and the pressure that is added there will make a gap. Thus the top lid of the containment vessel was pushed up by the pressure. Hence it allowed the hydrogen gas to leak. The TEPCO report says that there was some deterioration of the sealing material due to high temperatures, and I suppose that was possible. However, a slight amount of leakage through the seal cannot explain a huge amount of hydrogen gas discharge. I estimate that the lid was lifted up at least a few millimeters to allow the hydrogen gas to blow out from the periphery of the lid. In other words, the top lid was lifted up by the inner pressure.

Please look at Fig. 2.3. There is a small space of about 300 m³ called "reactor vault" above the containment vessel. On top of it, there is a large shield plug made of concrete which is removed in case of work such as refueling. The size of the plug is approximately 13 m in diameter and 2 m in thickness, and it weighs approximately 600 tons. A large crane is needed to remove it.

The hydrogen that leaked out by pushing up on the lid of the containment vessel blows out into the space of the reactor vault. Although the reactor vault space connects to a larger space beneath it, the communicating space between them is narrow so that it is difficult for the large volume of hydrogen that blew out sharply to be released into the lower space.

The reactor vault pressure rises with the hydrogen gas blowout. This is where many people fell into a pitfall in their thought processes.

What is the pitfall? It is the assumption that a 600-ton concrete plug will be as solid and immobile as the floor. Let's see if that is true by making a simple calculation. The specific weight of concrete is approximately 2.3 so that the weight per unit area of a concrete object with a thickness of 2 m is about same to that of a water column with a height of 5 m. This can be interpreted as only 0.05 MPa. In other words, if the reactor vault pressure exceeds 0.05 MPa, the pressure acting on the entire area of the bottom of the shield plug can lift this huge shield plug (ref. to Sect. 2.6.5).

Most of us think that the shield plug is so heavy that it is immobile, so that they simply assume that the hydrogen that entered into the reactor vault goes through a contact surface between the shield plug and the floor. This is wrong. The small amount of flow that could pass through a narrow gap in the contact surface cannot produce a hydrogen explosion. A gap of at least a few millimeters would be needed.

Rather, the pressure of the containment vessel's DW with a volume of 4,000 m^3 – the source of the pressure – is as much as 0.8 MPa. A very small amount of blowout of hydrogen from the containment vessel is enough to raise the reactor vault pressure to 0.05 MPa. I imagine that the reactor vault pressure rose to a value much higher than 0.05 MPa and stayed there for a while to keep the shield plug pushed up. During that period, the hydrogen gas from the containment vessel leaked out and spread to the entire fuel exchange floor. The flow stopped when the reactor vault pressure dropped and the force to raise the shield plug was lost.

Incidentally, as a result of the explosion that occurred in Unit 1 on March 12, the reactor building of Unit 2 had lost the blowout panel provided on the sealed wall of the fuel exchange room, thus leaving a big opening. Consequently, the hydrogen gas that blew out from the reactor vault leaked out to the outside of the reactor building through this opening. It was about 10 p.m., which coincided with the time when the DW pressure of the containment vessel indicated the maximum value (approximately 0.8 MPa).

Although it is only my personal view, I believe that hydrogen gas leaked out in this case most likely as a plume (columnar smoke) from a bonfire from the reactor vault to the outside via the blowout panel. As a consequence, hydrogen did not remain in the reactor building and that's why the building was saved from an explosion. The reason I thought of a plume for this case was that I knew that the hydrogen gas was very hot and hot smoke from a bonfire tends to flow as a plume.

Figure 2.4 shows the radiation level in the vicinity of the main entrance gate of Fukushima Daiichi immediately after the accident. Although the detail of this will be explained later, please note the changes from around 6 p.m. on March 14 through the end of March 15.

We notice two radiation level peaks at around 10 p.m. on March 14 and at 6 a.m. on March 15. Since there were no significant events during this period pertaining to Units 1 and 3, which had already had explosions, we can determine that these two peaks must be the radiation emissions from Unit 2.

The first peak occurred at 10 p.m. which coincides approximately with the core melt timing. As I explained before, the large amount of hydrogen gas that was generated simultaneously with the core melt leaked out to the fuel exchange floor by lifting the lid of the containment vessel and the shield plug and emitted out to the atmosphere through the blowout panel.

The second emission peak occurred at around 6:14 p.m. on March 15, which is noted on the TEPCO's report as an occasion of a sharp drop of the containment vessel pressure, dropping beyond the lowest measuring capability of the SC pressure indicator. I suppose it cannot be denied that the second peak was caused by the radioactive substances being discharged through the broken hole of the containment vessel. Please pay attention to the fact that it took 5–6 h for the DW pressure to come down. While the second emission was also a direct emission from the containment vessel, the radiation concentration is higher because of the progress of melting.

Both the first and second radiation emissions were direct emissions from the DW. Different from the cases of Unit 1 and 3, the radiation concentration was very high as it was not the radiation washed by the SC water. As it was not the discharge via the stack, the radioactivity was not diluted by the wind. It was discharged to the ground, so that the density was naturally higher. With these two radiation emissions from Unit 2, the radiation level surrounding the power plant rose and made the forced evacuation mandatory. I will be discussing this in Chap. 4 of Part II.

The containment vessel pressure after the first discharge was showing significant pressure readings in both the DW and the SC, so that it is assumed that no major damage was suffered in the containment vessel during this time period. However, after the second discharge, the DW pressure reading dropped extremely so that it is evident that the containment vessel was damaged. Therefore, it is estimated that the radiation emission was higher and the radiation dose value was higher the second time than the first time.

That concludes the explanation of the core melt and the radiation emission of Unit 2.

2.5.4 *Section Conclusion*

This has been a long explanation. Let me summarize at this point the core melt and the hydrogen gas emission in the case of Unit 2.

Although Unit 2 lost all electric power because of the tsunami, the RCIC pump, using decay heat, survived, cooling the reactor for 3 days. However, the RCIC pump finally was stopped by exhaustion at around 11 a.m. on March 14, and all means of cooling the reactor were lost.

Having lost all means of cooling, the reactor's decay heat was removed by water evaporation, so that the reactor's water level started to drop. As the water level dropped, the fuel rod temperature started to rise because of insufficient cooling. At 6 p.m. or so, the reactor water level dropped to the bottom of the core. However, the fuel rod temperature did not rise so much because the fuel rods were cooled due to the heat dissipation effect of breezy intermittent steam flows to the safety relief valve: I suppose it was below 1,000 °C, according to my ballpark calculation.

At about the same time, at 6:02 p.m., the safety relief valve was set and kept to the open position in order to lower the reactor pressure. As a result, high pressure steam continued to be released through the safety relief valve, and the reactor pressure dropped sharply from about 8 MPa to 0.4–0.5 MPa. As a result of the autonomous boiling caused by this pressure drop, the temperatures of the fuel rods dropped to around the water saturation temperature (150–160 °C).

If the seawater injection by the fire engine started at this point, the accident situation could have been substantially different. This is because a cooled cladding tube does not cause a zirconium-water reaction.

Unfortunately, the seawater injection by the fire engine started at 7:54 p.m., or 2 h later. During that time, the fuel rod temperature rose at least 1,500 °C.

It was approximately 1 h later, or around 9 p.m., when the injected seawater contacted the red-hot core. This delay was the time needed to fill the vacant space beneath the core, created by the decompression boiling, with seawater. The contact between high temperature zircaloy and water caused an oxidation reaction and generated a large amount of hydrogen gas. The reactor pressure rose sharply during this short period of time.

The oxide film that covered the hot cladding tube surface was quickly cooled, shrank, and broke apart, thus causing the fuel rods to break into pieces and the core to collapse. I am quite sure that a pile of debris of the collapsed fuel rods mounded on the core plate. However, since the water quantity was small, this reaction did not melt the entire core.

I believe the full-scale core melt occurred around 10 p.m. This delay was the time needed to replenish the seawater spent in the first reaction and for the seawater to start infiltrating into the agglomerates of debris. The contact between high temperature fuel debris and seawater initiated the zircaloy-water reaction. I am certain that this reaction started around 10 p.m. from the indicator of the reactor pressure gauge shown in Fig. 2.9, the DW pressure change among the containment vessel pressure changes shown in Fig. 2.11, and the sharp rise of the radiation dose shown in Fig. 2.4.

Heated by the light, hot hydrogen gas, the upper part of the containment vessel becomes hot. The bolts used for fastening the top lid elongated due to thermal expansion, thus reducing the force for fastening the top lid of the containment vessel, and the top lid is pushed up by the increased inner pressure. Hydrogen gas blows out through the gap formed as described above and fills the reactor vault space above the containment vessel. Since the reactor vault space is small, the pressure quickly rises and pushes up the shield plug on top of the vault. The shield plug thus lifts up and the hydrogen gas flows into the fuel exchange room through the gap formed by the lifting. Although it sounds like a long process, the time needed must have been relatively short.

It is conceived that, as a result of this first reaction, the bottom of the agglomerates of debris became partially molten and a thin egg-shell that serves as the bottom of the pot was gradually formed in the shape of a bottom of a pot. The inside of it must have been filled with a hot mixed molten substance consisting of uranium, zirconium and oxygen mixed in a non-uniform state.

The second and third pressure increases shown in Fig. 2.9 represent the reaction between the mixed molten substance and water caused by the seawater that flowed in across this pot-like boundary. It is imagined that the debris temperature rose even higher because of the combination of the reaction heat and the decay heat. That is the reason so that is why the second and third reactions were larger. That was the status of the core from midnight through 2 a.m. The reason that two pressure rises occurred was probably that the seawater supply from the fire engine was not large enough to cause the entire mixed molten substance to react all at once.

Because of the hydrogen gas generated in the second reaction, the containment vessel pressure jumped up to 0.8 MPa. That is twice as large as the design pressure. The hydrogen gas used for raising the pressure was generated from the molten core, so that it was very hot. It must have been somewhere near 2,000 °C, judging from the melting point of the mixed molten substance of uranium, zirconium and oxygen.

That is the outline of the core melt and hydrogen gas generation of Unit 2.

Although I realize that there are various estimates as to the molten core, I think that the core of Unit 2 still remains in the reactor pressure vessel judging from the curriculum vitae of the accident that water always existed at the bottom of the reactor pressure vessel. Another reason is that no significant change was observed on the reactor data since the late evening of March 14 when the molten core appeared.

The performance of the core melt is essentially the performance of a substance in a stewing pot. Since what was in the pot was the three-element mixed molten substance comprising of high temperature fuel and zirconium, each time water was added, a reaction took place, a thin skin was formed, uranium dioxide melted, and the molten part gradually grew in size. At the same time, the entire surface must have increased in thickness and became a hard shell.

Judging from the fact that the entire volume of the fuel rods is about half the core volume, I estimate that its final size is 3–4 m in diameter and 2 m in height and its shape is that of a semi-egg-shaped muffin.

I cannot find any data that shows that the molten core flowed outside of the reactor pressure vessel. It is not impossible that some amount of the mixed molten substance may have dropped through the cracks of the egg-shell but even if it dropped, it would have come into contact with the water below it and formed small balls. We have learned from the TMI accident that once the egg-shell is formed, it prevents any further chemical reaction from occurring. Therefore, we will probably find those small balls, cooled by the water beneath the core, having formed agglomerates of a certain kind of alloy and lying on the bottom of the reactor pressure vessel.

I presume that, if the molten core surrounded by the egg-shell still exists today as a mass of about 2 m high, the inside of it is still melted due to the decay heat although the exterior is cold. According to my calculation, the decay heat is still probably between 100 and 200 kW 3 years after the accident. However, it is heating inside a shell of a limited volume. It can be compared to having several hundreds of electric heaters turned on continuously in a tiny enclosed room of 4.5 *jo* (approximately 80 ft^2).

That will instantaneously cause a fire if that was a wooden Japanese house, but the egg-shell we are talking about is surrounded by thick walls made of three-element alloy. You can imagine a casting piece. The outside of the wall is cooled by cooling water but the core portion is still molten. The thick wall has cracks through which remaining radioactive refractory gaseous substances are leaking little by little. The radioactive substances are cooled by the cooling water so that most of them are solidified and deposited in the water at the bottom of the reactor pressure vessel. That is how I am imaging the status of the molten core at the moment (ref. to Fig. 2.13).

That concludes my summary explanation of Unit 2.

Let me list the new findings and their reasons from the Unit 2 accident in comparison with the findings from the TMI accident.

I. Neither a core collapse nor melting occurred for approximately 2 h until the water injection started, even though the fuel rods were without cooling and in a red-hot state because the water level dropped below the core. The heat under said condition was dissipated by radiation, which we experienced for the first time in a nuclear accident (Fig. 2.14). This will be discussed further in Sect. 2.7.

II. When the cool seawater was injected and reached the bottom of the core, the fuel rod splitting and the core collapse occurred, and the high-temperature zirconium caused oxidation by seawater, which in turn caused the core melt. This is the same process as the TMI core melt.

III. A large amount of hydrogen was generated as a byproduct of the core melt. A sharp rise of the containment vessel pressure thus developed resulted in the lifting of the top lid of the containment vessel, which allowed the hydrogen gas to leak via the gap produced by the lid lifting, which in turn allowed the hydrogen gas pressure to lift up the shield plug, thus allowing the hydrogen gas to enter the reactor building and leak to the outside environment via the blowout panel.

IV. As a consequence, radioactive substances from the molten core were directly emitted to the atmosphere, and the radiation level in the surrounding areas of the nuclear power plant rose to levels far exceeding the IAEA's evacuation advisory dose of 20 mSv/y.

V. It seems that Unit 2 would not have had a core melt if water was injected into the core just after the pressure was reduced.

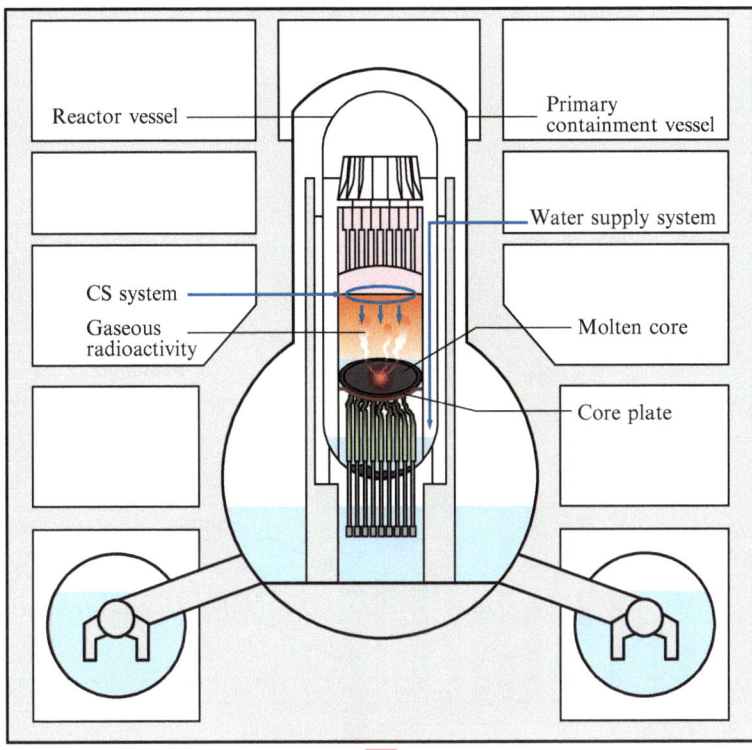

Reactor vessel

Primary
containment vessel

Water supply system

CS system

Gaseous
radioactivity

Molten core

Core plate

Enlarged view of molten core

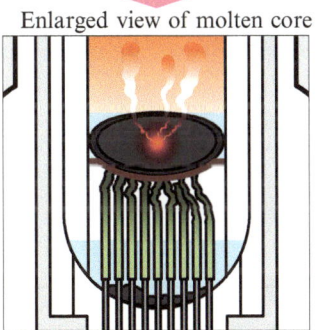

Fig. 2.13 Imaginary view of current status of molten core (Unit 2)

2.6   Case of Unit 3

Unit 3 is also a BWR of 784 MW capacity equipped with a MARK-I containment
vessel. There may be some differences in the details, but it can be regarded as a
sister reactor of Unit 2, as it is designed in a similar manner as far as the reactor and
the containment vessel are concerned.

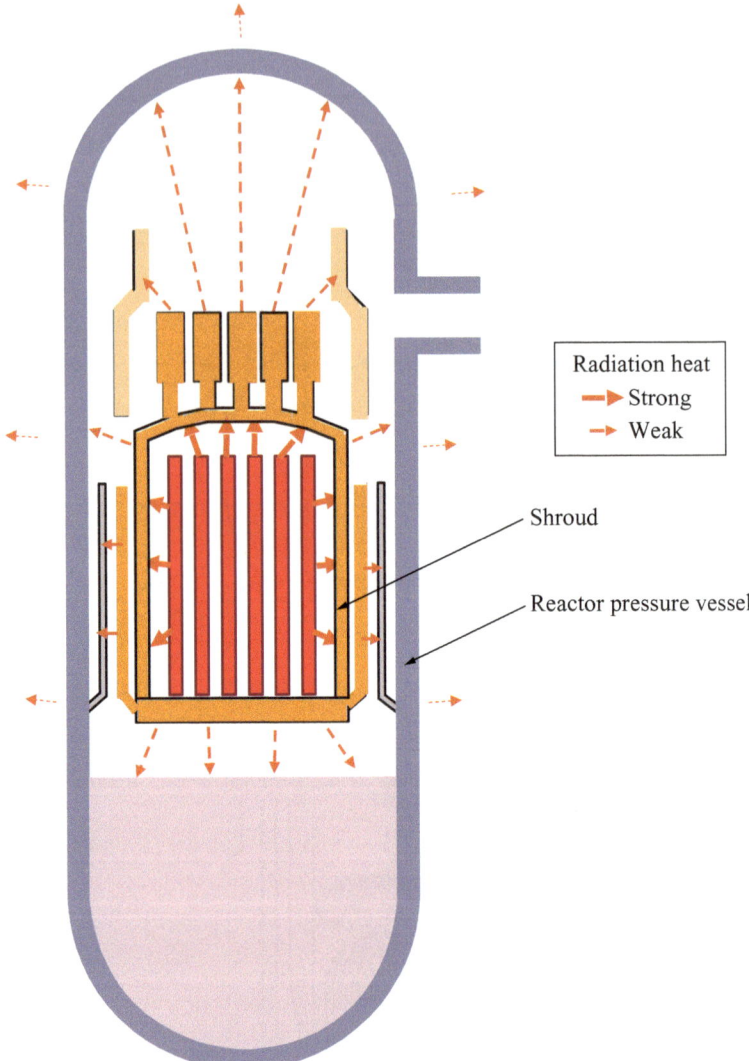

Fig. 2.14 Explanatory diagram of core heat dissipation by radiation

However, there is a major difference in the ways accidents occurred in Units 2 and 3. First, no explosion occurred in Unit 2, but the reactor building of Unit 3 was destroyed by an explosion. The containment vessel was damaged due to the failed venting in Unit 2, Unit 3 succeeded in venting and hence the containment vessel was kept intact. As a consequence, while that from Unit 2 was a direct release from the containment vessel, the radioactive emission from Unit 3 was released after being decontaminated by the SC water so that it was incomparably smaller relative to that of Unit 2.

It was lucky for Unit 3 that a portion of the DC power source survived the tsunami so that its RCIC could be controlled properly. Although things looked like they were going smoothly as the containment vessel vent was implemented as planned and the water injection by the fire engine was also working relatively smoothly, the core melted and a hydrogen explosion occurred in the reactor building at around 11:00 a.m., March 14.

Then why did the core of Unit 3, where everything seemed to be working smoothly, melt, resulting in an explosion earlier than that of Unit 2? Finding the answer to that question is the theme of this section.

To our luck, a portion of the DC power survived the tsunami, so that a certain amount of data during the accident is available to us. The data include the reactor pressure and the water level shown in Fig. 2.15 and the containment vessel pressure shown in Fig. 2.16. We will follow the path that led to the melt and the explosion based on these data. I will be asking you readers to look at these two charts again and again.

However, I must warn you that my explanation will tend to be lengthy as I have to deal with a lot of data available to us. The reason I do so is because I believe precise explanations of all the data is necessary in order to achieve the credibility of my explanation of what happened. However, I also realize that a voluminous explanation may make difficult to grasp the outline of the accident. Faced with this dilemma, I decided to explain the accident of Unit 3 by dividing it into the following five time periods:

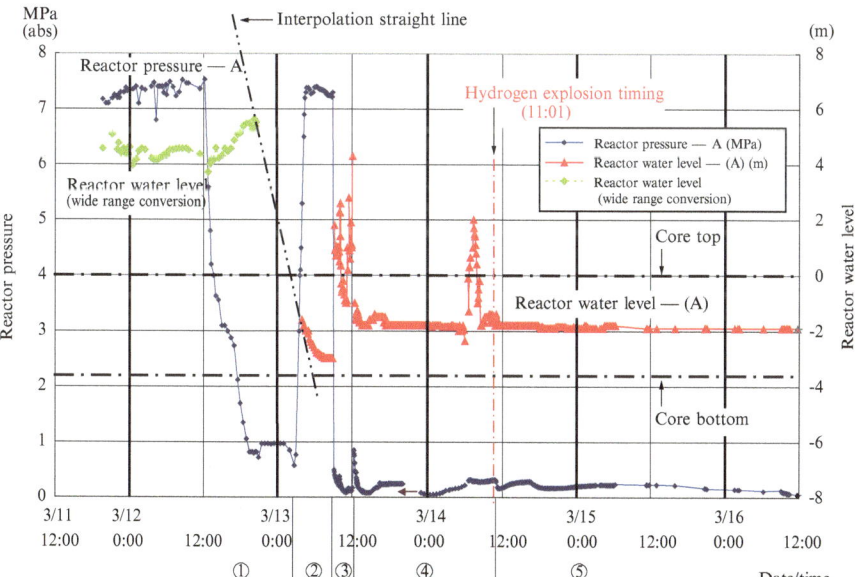

Fig. 2.15 Unit 3 reactor pressure and water level changes (measured data). Note: Time periods ① through ⑤ of the figure match with ① through ⑤ of 2.6.1 of the text (Source: from TEPCO "Fukushima Nuclear Accident Investigation Report")

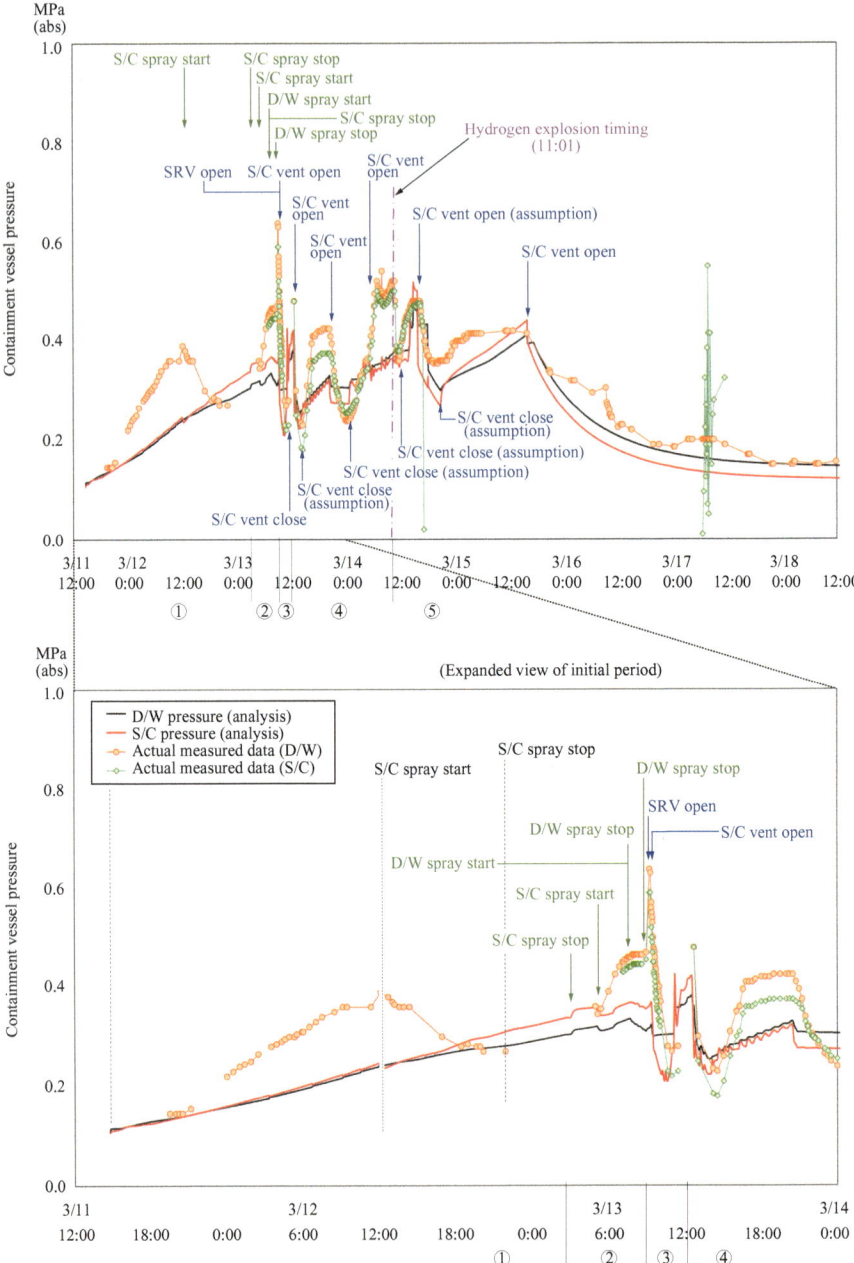

Fig. 2.16 Unit 3 containment vessel pressure change. Note: Time periods ① through ⑤ of the figure match with 2.6.5(5) of the text (Source: from TEPCO "Fukushima Nuclear Accident Investigation Report")

① Operating period of RCIC and HPCI
② Core temperature rise after HPCI stoppage
③ Start of core collapse
④ Seawater injection and core melt
⑤ Explosion of reactor building

At the end of the description of each period, we will confirm the state of the fuel rods and the core before entering into the explanation of the next phase. Please make sure to check the conclusion of each of Sects. 2.6.1 through 2.6.5.

As you read through, it is important for you to pay attention to the fuel rod (core) temperature, reactor water level, and cooling water injection status. The core melt occurs when the fuel rods in a red-hot state meet a large amount of cooling water, and hydrogen generation occurs simultaneously with the core melt.

Now is the time our up-hill battle begins.

2.6.1   Operating Period of RCIC and HPCI (① 4 p.m. on March 11–2:42 a.m. on March 13)

2.6.1.1   Operating Period of RCIC (4 p.m. on March 11 – Around Noon on March 12)

After losing electric power as a result of the tsunami attack, Unit 3 switched over to cooling via RCIC. This transition went smoothly as DC power was available.

Let us look at the data at the time of the accident of Unit 3. The reactor water level and pressure (Fig. 2.15) was kept completely stable until around noon on March 12. This is what makes Unit 3 different from Unit 2, and it was because the RCIC's flow adjustment was properly conducted as the DC power was alive. This part of the operation was beautifully done. When I first saw this data, I even jumped to conclusions that it must be an automatic operation.

The containment vessel pressure (Fig. 2.16) was rising slowly due to the steam release from the safety relief valve, etc., but it was quite a normal change. Everything looks fine.

However, at around noon on March 12, the all-important RCIC went into an automatic stop. The reason for that is not written in the TEPCO Report, and it is still written as "cause unknown" in the "Summary of Fukushima Nuclear Accident and Nuclear Safety Reform Plan" published by TEPCO now at the time of writing this manuscript. I am not happy about this lack of explanation.

I heard a rumor that the electric power was exhausted, but I wonder if that can be true. Since the power source of RCIC is the decay heat, the exhaustion of the DC power cannot explain the stopping of the pump. In fact, Unit 2's RCIC ran for 3 days without electric control, so that there is no reason why Unit 3's RCIC to stop only after 1 day. If it was caused by the electric circuit, it is a design fault. The cause analysis of such a minute point can lead to safety improvements in the future. This is what nuclear safety personnel should be aware of.

As RCIC stops, the water level of the reactor which lost water supply can only go down. As the reactor water level drops, the high pressure water injection system (HPCI) automatically kicks in. Although the reactor water level returns to a normal for a while, the reactor pressure begins to drop sharply as the water is supplied from the cold condensate storage tank. The rest will be described in the next sub-section.

Incidentally, the containment vessel spray from the SC started at around the same time as the stoppage of RCIC. This spray was stopped at around 2 a.m. of the following day. This is the reason why that the containment vessel pressure dropped during this time period. As this is not relevant to how the accident proceed, it is enough only to mentioned here.

Now, until about this time, i.e., around 11:30 a.m. on March 12, when the RCIC pump stopped, the fuel rods and the core were intact ((a) of Fig. 2.17).

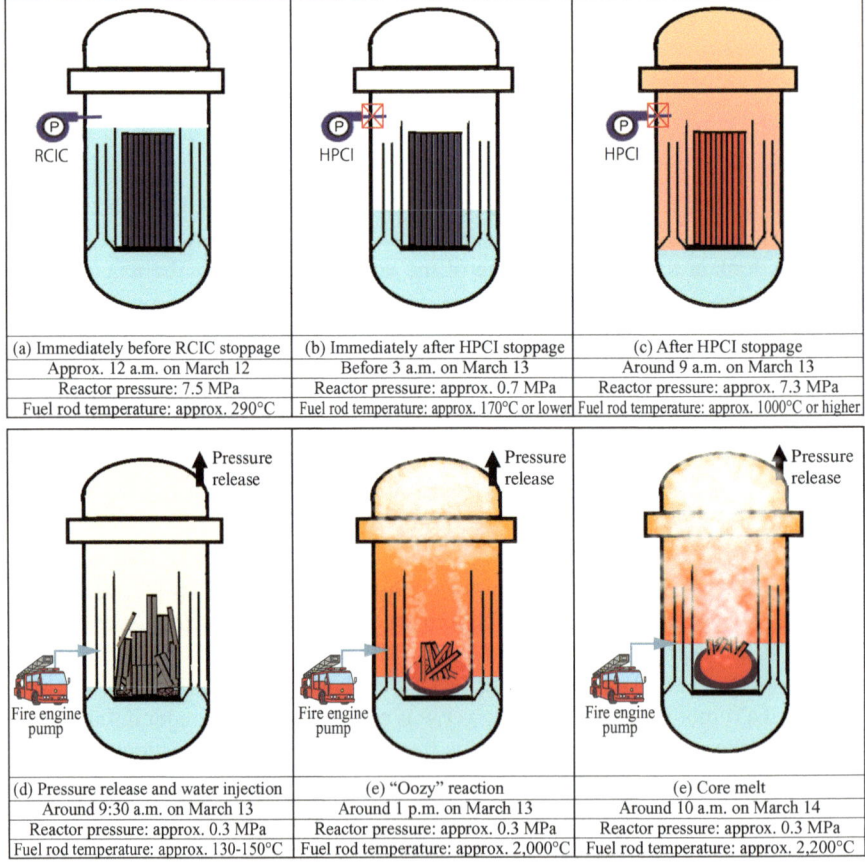

(a) Immediately before RCIC stoppage	(b) Immediately after HPCI stoppage	(c) After HPCI stoppage
Approx. 12 a.m. on March 12	Before 3 a.m. on March 13	Around 9 a.m. on March 13
Reactor pressure: 7.5 MPa	Reactor pressure: approx. 0.7 MPa	Reactor pressure: approx. 7.3 MPa
Fuel rod temperature: approx. 290°C	Fuel rod temperature: approx. 170°C or lower	Fuel rod temperature: approx. 1000°C or higher

(d) Pressure release and water injection	(e) "Oozy" reaction	(e) Core melt
Around 9:30 a.m. on March 13	Around 1 p.m. on March 13	Around 10 a.m. on March 14
Reactor pressure: approx. 0.3 MPa	Reactor pressure: approx. 0.3 MPa	Reactor pressure: approx. 0.3 MPa
Fuel rod temperature: approx. 130-150°C	Fuel rod temperature: approx. 2,000°C	Fuel rod temperature: approx. 2,200°C

Fig. 2.17 Progress of Unit 3 core status (model diagram)

2.6.1.2 Operating Period of HPCI (Around Noon on March 12–2:42 a.m. on March 13)

The advantage of Unit 3 was that its DC power survived. The core water level begins to drop sharply due to evaporation after the water injection cooling was stopped as a result of the RCIC stoppage. Receiving a warning signal of low core water level 1 h later, the high pressure coolant injection (HPCI) pump was turned on. It was 0:35 p.m. and it functioned properly as designed. The power source of this pump is also the decay heat.

The HPCI pump is designed to inject the cold water of the condensate storage tank – the temperature must have been approximately 20 °C – into the core directly. Unfortunately the capacity of the HPCI pump is so large that it overcooled the core. As a result, the reactor pressure dropped excessively, causing the HPCI driving steam pressure to drop, thus causing in turn the pump's speed to drop, and making the pump to be in such an unstable condition that it might stop at any time.

This situation can be easily understood in the variation of the reactor pressure as indicated in Fig. 2.15. The reactor pressure dropped sharply after noon on March 12, which was when the HPCI pump started to operate, and the pressure which used to be 7.5 MPa dropped to below 1 MPa by 6 p.m. By contrast, the reactor water level rose to its maximum of plus 6 m, almost reaching the top of the steam-water separator. Obviously this is an excessive cooling condition due to an excessive supply of cold water. The pump drive torque must have dropped substantially as a result.

The sharp drop of the reactor pressure accompanied with the HPCI operation caused a misreading of the reactor water level gauge. If boiling occurred in the reference condensing water chamber,[1] due to heat up by the ambient temperature rise, or when decompression was accompanied by the increase of the weight density of the reactor water, it will cause a false reading of approximately 1 m higher than the actual water level.

To put it more concretely, although the reactor water level reading at the time when the HPCI function stopped was approximately plus 7 m, the real reactor water level was, considering this error, approximately 6 m, or about the height of the top of the steam-water separator.

This misreading of the reactor water level gauge is described in (Appendix 2.6) which is at the end of this chapter. Please take a look if you are interested.

While the operating staff tried to reduce the water injection rate using a test line and the like, by around 3 a.m. of the following day the reactor pressure dropped to about 0.7 MPa, which is the lowest level that the pump can operate at. According to the TEPCO Report, the HPCI pump's speed dropped below the normal operating range. As a result of the lack of steam due to excessive cooling, the turbine for driving the pump was barely running.

Because of this sluggish running, the HPCI pump capability was essentially reduced to zero. Consequently, the reactor water level began to drop around 8 p.m. on March 12 and became minus 2 m (actual water level was minus 3 m) by the time

---

[1] A device for measuring the water level inside the reactor pressure vessel. It measures the pressure difference between the pressure vessel and the reference condensing water chamber.

the pump stopped at around 2:30 p.m. on March 13. Since the water level lowered 9 m in about 6 h, the lowering speed was 1.5 m/h. This information tells us a very important fact which will be described in the following subsection.

However, since the reactor pressure was lowered to around 1 MPa, it is safe for us to assume that both the reactor cooling water temperature and the fuel rod temperature were around 170–180 °C. Of course, both the fuel rods and the core were in a healthy state ((b) of Fig. 2.17).

2.6.2   Core Temperature Rise After HPCI Stoppage (② 2:42 a.m. on March 13–9:08 a.m. on March 13)

2.6.2.1   Reactor Pressure Drop

By this time, with no support from the outside available, the operating staff of the power plant seemed to have decided to solve the situation on their own. According to TEPCO report, they decided to make a last ditch effort by reducing the reactor pressure and injecting seawater using a fire engine. As the HPCI was gone, it was the only way to inject water.

In order to inject seawater using a fire engine, the reactor pressure needs to be lowered. This means that they needed to first lower the containment vessel pressure, which is located on the discharge side of the reactor. To inject as much water as possible to the reactor with such a low discharge pressure pump as a fire engine, it is necessary to lower the pressure injected as lower as possible. That is why not only the safety relief valve of the reactor but also the vent valve of the containment vessel has to be opened; that permitted the pressure from the reactor through to the containment vessel to be close to the atmospheric pressure.

It was explained that there are two vents, DW and SC, in the previous section. It was the SC vent which was used to make the gas in the containment vessel pass through the water in order to wash down radioactive substances.

Once they had decided to decrease reactor pressure and put it into practice, they hadn't felt any need to worry about the stop of the sluggishly operating HPCI. The site operating group went ahead with the task.

At 2:42 a.m. on March 13, the operating staff stopped the sluggishly operating pump.

The rest was to open the vent and go ahead with injecting water using the fire engine, but unfortunately it took some time to open the safety relief valve. Since they could not reduce the reactor pressure, they could not inject water. In the meanwhile, the reactor temperature and pressure started to rise again because of the decay heat.

At around 9:08 a.m. on March 13, the reactor pressure began to drop sharply as the safety relief valve was finally opened.

However it took as much as six and a half hours or so after the HPCI pump stopped before the pressure drop. Let us see what happened to the reactor in the meanwhile (Fig. 2.15). The reactor pressure rose again to 7.5 MPa and the reactor water level dropped to minus 3 m (in actuality minus 4 m). The core status changed substantially.

Incidentally, as we look at the water level data closely by interpolating the missing data with a straight line, we note that the lowering speed drops as the indication of water level gauge approaches minus 3 m from minus 2 m (in actuality minus 4 m from minus 3 m). Although it is only the data for a period of 1 h or so, it suggests that the water level does not change very much after the actual water level has reached the bottom of the core. This 1-h slowdown of the change in water level is very important.

This slowdown indicates that the heat removal from the core changed from heat conduction from the core to water to heat radiation to peripheral structures. As the core loses water, the decay heat is dissipated as radiation heat to the surrounding physical objects rather than heat used to evaporate water. Thus, the water evaporation rate decreased and the decrease in water level slowed down. By contrast, this caused the fuel rod temperature to rise. This radiation heat will be discussed in detail later in Sect. 2.7 of the chapter.

The slowdown in the decrease in water level means an increase of the fuel rod temperature.

In case of the foregoing Unit 2, the fuel rod temperature at the time when the reactor water level reached minus 4 m was estimated to be less than 1,000 °C at most (Sect. 2.5.2 of this chapter). The decay heat of Unit 2 was 0.4 %.

The condition of Unit 3 was not so much different either. Since it is estimated to be 2:30 a.m. on March 13 when the water level of Unit 3 dropped to minus 3 m, the decay heat of Unit 3 is estimated to be about 0.5 %, or slightly more than that of Unit 2. The core heat generation is about 10 MW and is capable of increasing the fuel rod temperature by about 1,000 °C/h.

Although it is a tough call to make, I say that the fuel rod temperature of Unit 3 had not reached a red-hot state, although it was slightly higher than that of Unit 2 and exceeded 1,000 °C at the time when the core lost water. The issue is, however, when did the water level lowering rate slowed down. I say that the less evaporation took place, the higher the fuel rod temperature rose, so that the center of the core must have reached a red-hot state. Please remember here the temperature rise of the fuel rods when they entered the sauna bath. The fuel rods had changed to the status [IV] of Fig. 1.5 ((c) of Fig. 2.17).

Leaving the discussion of the steps from here to the core melt to Sect. 2.6.3 of this chapter, let us review the steps from the stoppage of the HPCI pump to the pressure drop to see if there was any problem in the accident countermeasures. Let us take a few moments on this issue.

2.6.2.2   After Effects of the HPCI Pump Stoppage

Although the pressure reduction was finally achieved after a six-and-a-half hour struggle, this delay in reducing pressure was one of the factors that caused the core melt. More bluntly speaking, stopping the HPCI pump was the main cause of the failure. The core melt and the hydraulic explosion of Unit 3 would not have happened if the pump had not been stopped, but continued to operate sluggishly. I must explain here the reason why I believe so.

The evidence is Fig. 2.15. Please look at the reactor pressure data. Until 2:30 a.m. on March 13 when the HPCI pump stopped, the reactor pressure had been kept at 1 MPa or lower, which is very low. While this sluggish pump operation was in progress, there was no reactor pressure increase and the fuel rod temperature also remained low despite the fact that the core water level was decreasing because of evaporation. With only 0.5 % decay heat, the core could have been easily cooled by the breeze of steam and the fuel rod temperature must have been only up to 160–170 °C. What changed this state completely was the stopping of the HPCI pump.

As the HPCI was stopped, the reactor pressure started to rise instantaneously. It is because the steam that had been used for driving the HPCI lost somewhere to go to and the reactor became a sauna bath. The decay heat raised the fuel rod temperature while the steam that had no place to go started to raise the reactor pressure. This is the pressure increase shown in Fig. 2.15.

The heat generated by Unit 3 is estimated to have reached about 10 MW, assuming that the decay heat was slightly above 0.5 %, for a few hours prior to the pump stopping. This can evaporate approximately 25 tons of 7 MPa saturated water or 13 tons of seawater according to our calculations.

The next issue is the reactor water level lowering rate. The water level lowering rate during the period of 9 p.m. on March 12 through 3 a.m. of the following day was 9 m per 6 h. If this data is compared to the water level lowering rate of Unit 2, i.e., 10 m drop in 8 h (Fig. 2.8), we see that the two are relatively close. Considering the fact that the decay heat was approximately 0.5 % in Unit 3 and approximately 0.4 % in Unit 2, we can safely say that both are the same and are the results of the same cause.

The reason for the water level drop in Unit 2 was the reactor water's evaporation by decay heat. From this, we can safely assume that the reason for the water level drop in Unit 3 was the evaporation by decay heat, as in the case of Unit 2.

This has a very important meaning.

One is that it indicates that the HPCI pump was only running freely with no load, not pumping water at all, during the sluggish operation. If even a slightest amount of the cool water from the tank had been supplied, it would have affected the evaporation of the reactor water so that the water level lowering rate could have been different.

That means that the HPCI pump was not pumping water at this time, rather it was running idly, just like the idling of an automobile engine. It amounts to the fact that it was only serving the role of a conduit for transmitting the steam generated in the core by the decay heat to the SC via the turbine. In other words, it was serving as an outlet of the steam, or an outlet of the decay heat. When the operating staff stopped the HPCI pump, they closed this steam outlet themselves, thus taking away the means of removing the decay heat. As the reactor's water level lowering rate decreased, the reactor's temperature and pressure began to rise again.

If the HPCI pump continued this sluggish operation, the core would not have entered a sauna bath state, as it would have provided an outlet for the steam generated by the decay heat. The fuel rods were cooled by the breeze provided by the water evaporation, so that the fuel rod temperature would have never risen. This is the substance of the failure by the pump stoppage I raised earlier. The change in pressure and water level during the HPCI operation period are its proof.

In essence, stopping of the HPCI took away the outlet for the steam and caused the fuel rod temperature to rise.

I am not criticizing at all. I am simply suggesting that we should accept it as a fact. The fact that the sluggish operation of the HPCI pump was playing the role of a steam outlet would have been very difficult to think of at that time. I would not have thought of it, even if I was at site at that time.

As the outlet of the decay heat was closed, the reactor temperature and pressure naturally went up. That is the re-rising of the reactor pressure shown in Fig. 2.15.

Now, let us compare this to the reactor's condition in the TMI accident. Let's look at the reactor water level of Unit 3 shown in Fig. 2.15. Around 9 a.m. on March 13, just before the vent was opened, the status was such that the water level dropped to approximately minus 3 m (actual water level was minus 4 m), and the most of the fuel rods were exposed above the water surface – in other words, in a sauna bath state. In case of the TMI accident, the core was in a sauna bath state and the fuel rod temperature was rising after all of the reactor coolant pumps stopped and the block valve of the relief valve was closed. There was no current in the core after the HPCI pump was stopped in case of Unit 3 as well. In both cases, as time progressed the fuel rods went from being intact as shown in [I] to the red-hot state shown in [IV] of Fig. 1.5. Although there are some differences between the two in the water level, the stoppage of the HPCI created the red-hot condition of the fuel rods, similar to the TMI case.

If water comes in here, the core will collapse and melt. The lower half of the core was immersed in water in the case of the TMI accident, and contributed to the melting of the collapsed core. What did the injection of water into Unit 3, where the water level was below the core, bring about?

2.6.3   Start of Core Collapse (③ 9:08 a.m. – 0:20 p.m.)

At around 9:08 a.m. on March 13, the reactor pressure vessel's pressure began to drop sharply as the safety relief valve was finally opened. It must have caused decompression boiling of water that remained in the reactor pressure vessel and the bubbles generated must have caused the water level to rise temporarily from where it was below the core. The raised water suddenly cooled some of the red-hot fuel rods and then caused the rods to break up in pieces. Because of this breakup, the core collapse occurred in various places. As the extent of this collapse was a transient condition, we will not find it inside the core even if we are given a chance to look inside now, but I believe that the range of the core collapse was rather limited.

It does not make any difference to the final result. You are free to imagine though ((d) of Fig. 2.17).

Please pay attention to the change in the containment vessel pressure after the water injection was started (Fig. 2.16). The graph is rather complicated with peaks and valleys but the DW and SC data show matching tendencies so that we can trust the data.

The containment vessel pressure rose in one stroke up to about 0.6 MPa as the safety relief valve was opened, and dropped to about 0.3 MPa later as a result of the containment vessel venting (Fig. 2.16). On the other hand, it looks like the reactor pressure also dropped steeply (Fig. 2.15), and reached around 0.2–0.3 MPa in one stroke. However, since there is no way that the containment vessel pressure can be higher than the reactor vessel, I suppose both of them were around the same pressure of 0.5–0.6 MPa.

Immediately after the pressure releasing, or around 9:26 a.m. on March 13, the water injection using the fire engine started. With this water injection, I assume that some amount of reaction occurred between the collapsed fuel rods and water. My assumption, however, is based on these sharp changes of the reactor water level (Fig. 2.15) and the containment vessel pressure (Fig. 2.16) observed in the data after the water injection, and it is not a confirmed conclusion. This reaction was a localized one, not one that could develop into an overall core melt. This is because there was not enough water.

The amount of water injected into the core during this period between 9:25 a.m. on March 13 and the explosion on March 14 is not determined exactly. Since the water that a fire engine can pump is about 25–40 tons/h, the amount of water that should remain in the reactor must be approximately 10–25 tons at most after subtracting the evaporation of approximately 13 tons due to the decay heat from the above amount and also considering the fluid resistance of the hose and other piping connected. According to this calculation, the reactor water level is supposed to rise, but the water level shown in Fig. 2.15 is different. It shows only a constant value.

The water level shown by Fig. 2.15 goes up and down wildly for a period of about 3 h from 9:25 a.m. through 0:20 p.m. suggesting the occurrence of a zirconium-water reaction. After ups and downs, the water level finally settles down at minus 1 m. The increase of water volume from the time when the water level was minus 3 m can be calculated as about 12 tons. In other words, the increased amount of water per hour is approximately 4 tons, so that the calculation does not match unless approximately 8–23 tons of water evaporated per hour for some reason. This evaporation amount is too much even if the water evaporation due to the zirconium-water reaction is estimated at the maximum.

More problematic is the amount of water injected for a long period of time from 1:12 p.m. on March 13, when the water injection is restarted, until 11 a.m. on March 14, when the explosion occurred. If we assume that the pump was operating properly, it makes no sense at all that the water level stayed constant at 2 meters constant as shown in Fig. 2.15.

Since I could not solve this dilemma, I decided to explain the core condition disregarding the water amount calculated based on the fire pump capacity, and instead assumed the measured water level data is correct. If the pump was operating properly, the explosion of Unit 3 would not wait until 11 a.m. on March 14. However, this was proven to be a correct answer as explained below.

It was on December 13, 2013, when I was finishing up the final draft of this book, that TEPCO published a report of its investigation and examination of unidentified and unsolved matters. According to the report, there were several branch pipes connected to the fire engine pump so that the water injected into the core was not the same as the pump outlet capacity, but rather less than what was anticipated. In other words, a portion of the water from the pump was diverted elsewhere. This publication answered all the questions I had concerning the water injection using the fire engine. I learned that the water injection was done in accordance with the containment vessel pressure by automatically diverting the excess amount to the outside.

I also learned from the TECPCO report that the delivery capacity of the fire engine pump that they used was 75 tons/h. It was much larger than what I assumed in my analysis for writing this book, which was 25–40 tons/h. I decided that I would not rewrite the draft however. I decided so because the discrepancy is irrelevant to the explanation of the core situation if the case was as described above.

I suppose that the zircaloy-water reactions continued to occur intermittently between the water, which kept gradually rising as seawater was injected, and the agglomerates of debris for a few hours until 0:20 p.m.. I also believe that an egg-shell similar to the one observed in the TMI case was gradually being formed on the contact boundary between the water/steam and the molten core.

This core melt condition caused by water shortage, which was also described for Unit 2, occurred earlier in Unit 3, chronologically speaking. The zirconium-water reaction that occurred for a short period of time in TMI is not necessarily the same as what occurred in Units 2 and 3 under the water shortage conditions.

The random fluctuations of the containment vessel pressure data until around noon of that day seem to indicate the operators' intense efforts to combat the core reactions. According to the TEPCO report, they did everything they could manage to keep the vent valve open, including replacing the gas cylinder and temporarily installing an air compressor. They also manipulated the containment vessel spray, opening and closing it, in order to reduce the containment vessel pressure. All these manipulations were intended to lower the containment vessel pressure that is generated by the water evaporation due to the decay heat and the reaction heat. I have to remind you that there was no electric power at the site. They had to do all of these in the pitch dark. They did all they could do.

2.6.4   Sea Water Injection and Core Melt (④ Approximately 22 h from 0:20 p.m. on March 13 Through 11 a.m. on March 14)

At 0:20 p.m. on March 13, the water source was switched to seawater as the water supply from the fire-protection tank was running low. The water injection was interrupted during this switching period of about 1 h. The TEPCO report says that it was 1:12 p.m. when the seawater injection started.

What happened in the core during this 1 h period is unknown. I can only guess as no data is available but I suppose no substantial change occurred. I assume that the core water evaporated during this period because of the decay heat and the containment vessel pressure was rising slightly. During this period, the DW spray of the containment vessel was operating to lower the pressure.

During the period of about 22 h from 0:20 p.m. on March 13 through 11:00 a.m. on March 14 when the explosion occurred, the data of Unit 3 show no significant change except some fluctuations of the containment vessel pressure.

The reactor pressure (Fig. 2.15) maintained a constant value of approximately 0.4–0.5 MPa. The indicated reactor water level maintained an approximately flat value of minus 2 m (actual water level 3 m). This gives us the illusion that it is a stable condition. The containment vessel pressure change shown in Fig. 2.16 is simply going up and down in accordance with the aforementioned opening and closing of the vent. There was no significant change that needs to be described. It was like the calmness before a storm.

There was a precursor of the storm however. It is the fact that there were several sharp increases of the reactor water gauge reading starting around 7 a.m., just before the explosion (Fig. 2.15). The abrupt change of the reactor water gauge reading is the evidence that there was some kind of a significant change to the reactor. Please remember this precursory event.

The operation record on the TEPCO report simply mentions that there were vent opening/closing operations as shown in Fig. 2.16 and also that the water injection by the fire engine was interrupted at 1:10 a.m. on March 14, in order to replenish seawater to the injection pit, but the water injection was restarted at 3:20 a.m. However, it shows that the seawater injection was interrupted for 2 h. This is a key point. Please remember this as well.

It is difficult to provide a definitive description for this status for which the only data available to be used as the basis of analysis is the containment vessel pressure, but it is safe to say at least that there was an insufficient amount of seawater to cause a change to the core.

Although I am not sure if it is an appropriate example to tell you, I experienced a situation that reminds me of these long, drawn-out hours [of indecisive situation].

About 20 years ago, My student spilled an amount of high temperature liquid sodium on an iron tray during an experiment by mistake. Metallic sodium is a material that can be oxidized much more easily than zircaloy and causes a violent oxidation reaction and explosion if it is placed in water even in an environment of a normal temperature. Metallic sodium also forms an oxide protective film by catching moisture in the air and burns oozily as it gets slightly hotter.

The relatively large amount metallic sodium spilled on the iron tray continued to burn oozily for a long period of time. The surface of oxide film keeps inflating and bursting repeatedly, the film constantly moving as it inflates, and flickering red and blue small flames can be observed through the cracks. It looked just like a hot pool of mud I once saw in the Beppu Hot Spring Spa. The burning continued almost endlessly.

I am imagining that the state of the core that existed almost one full day from 1:12 p.m. on March 13 through 11 a.m. on March 14, was a minor zirconium-water reaction similar to the oozy burning (reaction) condition I once saw. There is no concrete evidence to claim it, but I don't know how else I can explain the state that continued for almost 20 h while causing no significant event.

I am not certain about the amount of water injected by the fire engine, but I imagine that it was slightly more than the amount of seawater evaporated, or 13 tons/h. If we assume a certain fraction of the amount of water is used for the oozy burning reaction between zirconium and water, we can see some matching stories here. The combination of the heat from the oozy burning and the decay heat was used for evaporating the water and raising the temperature of the mixed molten substance (core) little by little. Moreover, I suppose that the hydrogen gas thus generated and the water vapor together contributed to raising the containment vessel pressure and necessitated the operation of the vent.

That was the explanation of the indecisive condition that continued almost a full day ((e) of Fig. 2.17). The precursory phenomenon of the abrupt water level jump that occurred around 10 a.m. on March 14 was an indication of increasing water to reach more actively with the hotter core.

As we think of the existence of this oozy reaction, we can see clearly that the core collapse and melting were entirely different things. Come to think of it, while there were three major water level fluctuations (reactions) in the core melt of Unit 2 described in the previous section, we can understand the phenomenon better if we see that it was the oozy burning time, waiting for more water.

I would like to leave the verification of this finding related to the zirconium-water reaction under a water shortage condition for studies by future scholars (Appendix 2.7). I also hope that they use the result of the verification for the future safety of nuclear engineering.

At 1:10, a.m. on March 14, the water injection using the fire engine was interrupted for filling up the pit that had been used as the reservoir of the seawater to be injected. It was 3:20 a.m. when the water injection was restarted. The seawater injection was interrupted for as long as 2 h. This is the operation I asked you to remember above, but this was the beginning of the disaster.

During this 2 h period, the water that contributed to the oozy reaction with the core disappeared by evaporation (Fig. 2.15), and the mixed molten substance (core) that became like a sauna bath was probably close to the melting point because of the decay heat. The structural members surrounding the core must have partially melted down by the radiation heat similar to the case of Units 1 and 2. It is assumed that, most likely, a portion of them was melted and mixed with the debris to form a certain alloy. And, that must have included such items as boron carbide (B_4C), which is used for the control rods of BWR.

With the seawater injection restarted at 3:20 a.m., the mixed molten substance mixed with this high temperature alloy came to be gradually immersed into the seawater. The timing of the contact between the seawater and the mixed molten substance is estimated to be around 7 a.m. judging from the water level fluctuation found in Fig. 2.15. Although there is a time difference of about 4 h between 3:20

a.m. when the water injection started and this time (7:00 a.m.), it must be the time needed to replenish the amount of water that evaporated during the water injection stoppage. A chemical reaction starts when water contacts with the sufficiently hotter core. It probably must have generated hydrogen gas and formed the egg-shell as well. It was probably because of the hydrogen gas generation that the containment vessel pressure of about 0.5 MPa was maintained for 4 h from approximately 7 a.m. until the explosion (Fig. 2.16) although the vent was open.

The reactor water level jumped up abruptly to plus 2 m about 1 h prior to 11:01, a.m. (around 9–10 a.m. according to Fig. 2.15), when the hydrogen explosion occurred in the reactor building of Unit 3. This is the precursory phenomenon I asked you to remember.

A violent reaction most likely must have started at this time between the injected seawater and the mixed molten substance that was partially protected by the egg-shell. I believe that it was started when the seawater completed the invasion of the mixed molten substance which had been protected by the egg-shell, so that the skin was broken by the inner pressure increase of the egg-shell, or the seawater started to react violently with the zirconium mixed in the mixed molten substance across the egg-shell. I suppose that the sudden change of the water level gauge data indicates the water level inflated by bubbles when a large amount of hydrogen was developed by the reaction.

With this reaction, the core started to melt throughout the reactor. The reaction was a violent one, although it might have been intermittent so long as the speed of water injection could catch up the reaction.

Around 10 p.m. on March 14, the Unit 3 core melted. It goes without saying that a large amount of hydrogen gas was generated simultaneously ((f) of Fig. 2.17).

The progressive steps of the core melt and hydrogen generation were the same as described earlier. Although it probably did not have a big effect, it is possible that the salts in the seawater may have affected the reaction chemically. This needs to be studied in the future.

2.6.5 Explosion of Reactor Building (⑤ 11:01 a.m. on March 14)

Let us check here the change of the containment vessel pressure (Fig. 2.16) up to about 11 a.m. when the reactor building explosion occurred. Although the vent operation further complicates the matter, it is clear that the containment vessel pressure that had stayed flat around 0.5 MPa until 11 a.m., which is estimated to be the time of the reactor building explosion, dropped instantaneously to 0.3 MPa, and rose again sharply in the next instance. This sharp drop of the pressure is the time when the hydrogen explosion occurred.

This sharp pressure drop is a temporary drop caused by the release of hydrogen gas to the reactor vault as the top lid of the containment vessel that had been heated by the hydrogen gas was lifted up. The released hydrogen gas entered the reactor vault and pushed up the shield plug to go out to the fuel exchange floor. The mechanism of the containment vessel hydrogen gas escaping to the fuel exchange floor is the same as that of Unit 2 explained in Sect. 2.5 of this chapter.

In case of Unit 2, the hydrogen gas escaped into the atmosphere as the blowout panel of the fuel exchange room had already been dislodged. However, the panel was still in place in case of Unit 3. The reactor building still maintained a semisealed condition. The hydrogen gas which had nowhere else to go must have risen to and flowed along the ceiling, changed its direction, agitated the room air, circulated, and spread throughout the room. The hydrogen gas that mixed with air became diluted and thus became an explosive gas. Incidentally, the deflagration/detonation range of hydrogen in air in terms of its volumetric ratio is 4–75 % – in other words, very wide.

Hydrogen that has turned into an explosive gas can ignite instantly if it finds an ignition source. There is a wide variety of ignition sources, including sparks, electricity, and physical impact. However, since the power station was out of electricity then, an electrical ignition source is out of question. A physical impact was the ignition source in this case. It was an impact caused when the shield plug, which was lifted by the pressure as explained in case of Case 2, dropped.

The lifted shield plug cannot stay up in the air forever. When the gas pressure that supported the plug is gone, the plug drops due to gravity. When the heavy plug drops and hits the concrete floor, it must have generated a spark as a flint stone would. Maybe no spark was generated but the sound of impact must have been quite impressive. That is enough to ignite the gas. The time when the shield plug landed was 11:01 a.m., i.e., the time the explosion of the reactor building of Unit 3 occurred.

I am sure many of you wonder if such a heavy shield plug can be lifted up. Let me show you an example. There is an actual example where a much heavier plug was lifted.

Figure 2.18 is a sketch made at the Chernobyl accident site. A large circular disc can be seen sitting in a vertical position in the upper part of the hollowed-out core. This is the shield plug of the Chernobyl reactor, which is approximately 13 m in diameter and weighs 1,600 tons. It weighs almost three times that of the BWR shield plug, which weighs 600 tons.

The reason that it is vertically positioned is that the shield plug was lifted up by the hydrogen gas pressure when the explosion occurred in an obliquely upward direction relative to the original position of the plug. The shield plug made a threequarter-turn before it landed just as if a judo player is attacked by a foot sweep just as he jumped up. You can tell that it was lifted fairly high from the fact that this large disc was rotated almost full turn.

The beard-like lines shown attached to the shield plug are the pressure tubes which contained fuel assemblies. The pressure tubes were built in such a manner as to extend through the core and to be attached to the shield plug. The hydrogen pressure that applied to the bottom of the shield plug was strong enough to tear off these pressure tubes to lift the shield plug up in the air. The investigators of the former

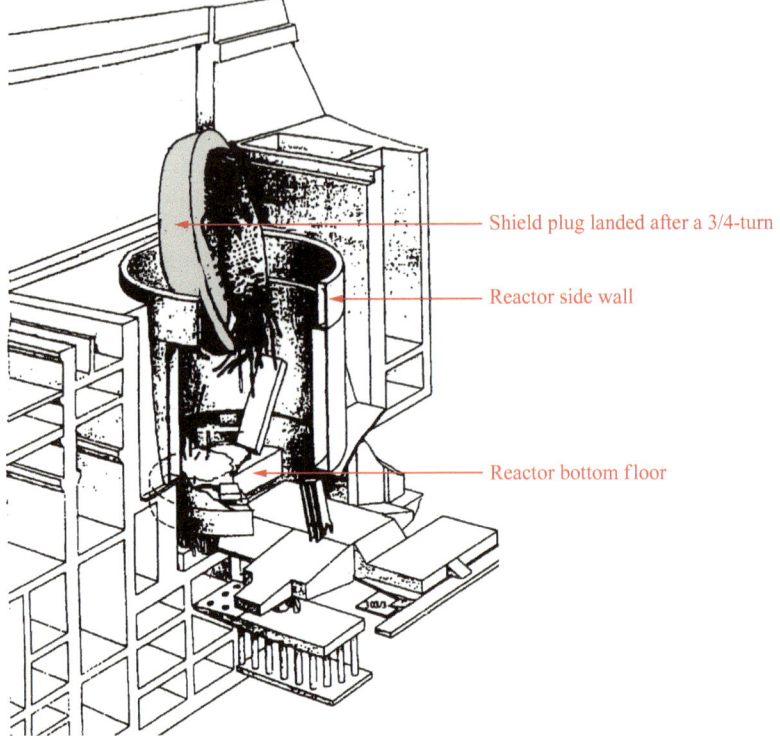

Shield plug landed after a 3/4-turn

Reactor side wall

Reactor bottom floor

Fig. 2.18 Chernobyl reactor's appearance after the accident [1]

USSR estimated that the total pressure that applied to the bottom of the shield plug was approximately 1 MPa.

As this sketch from the Chernobyl accident indicates, the shield plug can lift up if the gas pressure applies to the bottom of the shield plug.

It is reported that the ignition source in the case of the Chernobyl accident was the zirconium oxide debris from the hot fuel cladding tube. As the tubes built through the core were torn off, the fragments of the red-hot fuel cladding tubes were scattered all around, mixing with the steam emitted from the tubes. One of those pieces ignited the hydrogen gas mixed with air and caused the explosion. The explosion was caused when the shield plug was high up in the air. It is different from the impact of the drop of the plug, as in the case of the Unit 3 explosion. Otherwise, the shield plug of the Chernobyl reactor could not have made the three-quarter-turn.

Lifting the shield plug of Unit 3 was pretty easy compared to lifting the shield plug of the Chernobyl plant. As we calculated it in case of Unit 2, it lifts up at least theoretically if a pressure of 0.05 MPa or so is applied.

Then why didn't an explosion occur in the case of Unit 2 due to the impact of the shield plug dropping, when the shield plug was lifted up in a similar manner? It is because the hydrogen gas did not become an explosive gas through mixing with air, as in the case of Unit 3. In the case of Unit 2, the hydrogen gas escaped from the reactor vault into the air as a plume. The hydrogen gas density of this plume is approximately 100 %. A plume that is as hot as 2,000 °C flows in clumps just like water in a brook, and neither agitates air nor mixes with it. Since it is a flow of non-ignitable 100 % hydrogen gas, it does not ignite even if there is an impact.

I must explain further the hydrogen gas of Unit 3. As it is clear from TEPCO's investigation, the cause of the Unit 4 explosion was the hydrogen gas from Unit 3. How was it possible that enough hydrogen gas to cause an explosion entered Unit 4?

Let me explain in the simplest manner avoiding details. Unit 3 and Unit 4 used the same stack (chimney) to discharge the exhaust from the standby gas treatment systems (SGTS), which are reactor air-conditioning facilities – The units shared the stack. The tail ends of SGTS are connected to the bottom of the stack and are open upward. It has been concluded that the hydrogen gas of Unit 3 flowed backward via the connecting part through the SGTS of Unit 4 and entered the reactor building of Unit 4.

Although I cannot go into detail here, I believe it suffices to say that this backward flow theory was proven by the fact that the contamination of the filter for removal of radioactive substances placed in the SGTS of Unit 4 was heavier on the stack side outlet and lighter on the building side. If the exhaust gas does not flow backward, no contamination from the outside to the inside can occur. In other words, it was like the smell from your neighbor's kitchen being detected in your apartment's kitchen.

So we know the entrance path of the hydrogen gas, but the more difficult question is how enough gas as to cause a hydrogen explosion could flow backward. In order to explain this possibility, we must think about the pressure difference and the flow time.

We can see that the pressure difference can be relatively high if we consider the fact that the hydrogen gas that entered the Unit 4 building was the gas that flowed into the bottom of the stack via the containment vessel vent. The pressure at the vent outlet is the pressure difference in comparison with the pressure inside the Unit 4 building, which was at atmospheric pressure. With sufficient time, hydrogen gas can flow into the Unit 4 building little by little.

The question is the time such flow took place. The first thing that comes into our mind is the pressure of the hydrogen gas generated by the violent zirconium-water reaction that lifted the shield plug. Although this pressure is very large, the reaction does not last long. Although it is possible that a portion of that gas flowed into Unit 4, it is not enough to cause the explosion.

The next thing that comes into our mind is the gas from the vent. The time period the vent was open after Unit 3 caused the core collapse and started to generate hydrogen was approximately 2 h from around 8 p.m. on March 13 and approximately 2 h from around 9 a.m. on March 14. A total of 4 h – that is one hint. Since the containment vessel pressure ranged from approximately 0.2 to 0.5 MPa, that was also sufficient.

Although earlier I spoke unkindly about the containment vessel pressure change during this time, saying that it does not provide us any hints as it only coincides with the vent opening/closing, I have to retract that. It contained an important hint for solving a quite significant event, i.e., the explosion of Unit 4.

The more probable cause was the hydrogen gas generated by the oozy zirconium-water reaction. This is a long running reaction. Although it is not accurate, there is a possibility that it could have run for 26 h from around 9 a.m. on March 13, when the vent was opened, until around 11 a.m. on March 14, when the explosion occurred. During this period of time, hydrogen could have moved little by little to Unit 4.

Those of you who have very watchful minds may wonder if the vent was open during the entire time the oozy reaction was ongoing. Actually, there is no need for it to be that way. Let us think, just for the sake of argument, that the valve is closed.

The decay heat is constantly generating steam. Since this steam is blown out to the containment vessel, the containment vessel pressure keeps rising constantly. So long as there is the decay heat, the pressure should be rising constantly. On the contrary however, we see that there was a time when the pressure stayed unchanged; for example, it stayed around 0.4 MPa for 3–4 h from 5 p.m. on March 13 (Fig. 2.16). That is the hint.

It is interesting to note that the containment vessel pressure rose sharply just before it went into a standstill at 0.4–0.5 MPa. This phenomenon can only be explained by assuming that a certain amount of gas leaked from the containment vessel. We can guess that the hydrogen gas generated by the oozy reaction probably leaked out from the containment vessel to the reactor building.

Then what was the opening that allowed this leakage? Although we cannot determine for sure, the places where the opening can occur is either a gap produced at the top lid of the containment vessel, the hatches through which people or things pass through, or the penetration areas where cables or pipes go through.

I have heard that someone talked about leakage found in an equipment entrance hatch in Unit 3 through which various items were transferred. The door to the equipment hatch of the containment vessel is sealed by dual gaskets. If the seals are broken by internal pressure, it could well develop into an opening. Incidentally, 0.4–0.5 MPa is about equal to the design pressure of the containment vessel. The door seal is one of the most likely suspects. Of the potential leakage points, I think that the equipment transport hatch was the dominant one among them.

I believe that the hydrogen that leaked into Unit 4 is the hydrogen generated by the oozy reaction. The reason can be found in the explosion condition of Unit 3. The explosion of Unit 3 occurred in the fuel exchange room (refer to Fig. 2.3), which was immediately followed by the explosion in the mezzanine, which ended up damaging the building seriously. TV pictures showed the blast reaching several hundred meters high into the sky. Unless a large amount of hydrogen gas was residing in the lower level of the reactor building, such a huge vertical explosion could not have happened.

Hydrogen tends to move up as it is light. There is no physical reason for hydrogen gas to climb down the narrow staircase to the lower level. The reason that there was such a large amount of hydrogen gas as to cause an explosion was that the leakage was occurring on the lower level. The equipment transport hatch is located on the first floor of the reactor building. If we assume that hydrogen was leaking out for a long time through it, the explosion of Unit 3 can be explained beautifully.

I thus imagine that the hydrogen gas that flowed into the reactor building flowed further into the lower part of the stack via the duct of the standby gas treatment system and flowed backward little by little over a long period of time into Unit 4.

It seems to me that both a strong flow while the vent was opened and the chronic flow while the vent was closed took place.

I realize that my explanation on the hydrogen gas flow into Unit 4 extended into the explanation of the state of the explosion in Unit 3. Let us get back to the explosion of Unit 3 to summarize it.

As the injection of cold water using the fire engine started at around 9 a.m. on March 13, the fuel rods were broken into pieces, thus starting the core collapse. At the same time, a small amount of zirconium-water (most likely steam vapor) reaction started. This oozy reaction continued for a long time without stopping. As a result of the steam generation due to the heating by the reaction as well as the decay heat, the containment vessel pressure kept rising. The operating staff operated the vent valve intermittently in order to release the containment vessel pressure.

At around 10 a.m. on March 14, a violent core melt occurred because of the zirconium-water reaction. The timing can be identified by the sharp increase in the reactor water level. This reaction caused a large amount of hydrogen gas to be released into the containment vessel. This hydrogen gas was very hot. The upper structure of the containment vessel was heated by the hot gas, which in turn caused a gap in the top lid because of the elongation of the fastening bolts, and caused the hydrogen gas to be leaked to the reactor vault. From this point on, the sequence of events is the same as in Unit 2. The hydrogen gas leaked to the reactor vault pushed up the shield plug and went into the fuel exchange room on the fifth floor of the reactor building.

Different from Unit 2, the fuel exchange room was held in an air-tight condition. The hydrogen that flowed into the fuel exchange room swirled around the room, became an explosive gas, ignited due to the impact of the shield plug dropping, and caused the explosion. Because of this explosion, the upper portion of the fuel exchange room of Unit 3 was blown off. Almost immediately afterwards, another explosion was caused by the hydrogen gas which had crept into the bottom of the building and the reactor building of Unit 3 was seriously demolished, with the blast reaching several hundreds meters into the sky.

That is the summary of the core melt and the reactor building explosion of Unit 3.

2.6.6   Section Conclusion

As I had to deal with a lot of data, this was quite a long explanation. I tried to explain things in a way that is easy for people to understand, but the phenomenon was quite complex. Let me try a general summary here.

Although Unit 3 lost all sources of AC power, some DC power survived so that the core was kept cool while the RCIC and HPCI were operating. Had power been restored at this stage, no core melting would have occurred in Unit 3.

However, as the power plant lost electricity and experienced a water shortage, they decided to try to cool the reactor by forcibly reducing the reactor pressure and injecting water with the help of the fire engine. During this process, the core became red-hot and collapsed when the core water temperature dropped at around 9 a.m. on March 13 due to the decompression boiling. Had this decompression operation been conducted without stopping the HPCI, Unit 3 could have been saved from the core collapse and hence the core melt.

After this core collapse in the morning of March 13, both the reactor water level and the reactor pressure were relatively stable as if to keep balance with the water injection amount. However, the containment vessel pressure alone kept rising. Things looked rather stable until 9 a.m. on March 14. During this period, the zirconium-water reaction was continuing sluggishly in proportion to the amount of water supply. This was the oozy reaction, the prelude to the core melt. Although the containment vessel pressure looked as if it was being tamed under the operating staff's vent valve control, in reality when the pressure reached 0.4–0.5 MPa a gap developed in the equipment transport hatch of the containment vessel, and hydrogen gas leaked out through it. This hydrogen gas infiltrated Unit 4 next door and became the cause of its explosion.

At around 10 a.m. on March 14, because of the water level that had been gradually rising for almost a full day, a violent zirconium-water reaction started in the Unit 3 core, causing the core to melt and generating a large amount of hydrogen gas. This hydrogen gas pushed up the top lid of the containment vessel, flowed into the fuel exchange room on the fifth floor of the reactor building, mixed with air to become an explosive gas, ignited due to the impact of the landing of the shield plug, and caused the explosion. It was 11:01 a.m. on March 14.

That is the summary of the core melt and the explosion of Unit 3.

The following is the summary of the important conclusions concerning the melting and explosion of Unit 3:

I. The cause of the core melt is, same as in Unit 2, the violent reaction between the mixed molten substance (mainly zircaloy) and water.

II. A large amount of hydrogen gas generated by the violent reaction lifted the top lid of the containment vessel and the shield plug, flowed into the reactor building, mixed with the air in the room, and caused the explosion in the reactor building of Unit 3.

III. The ignition source of the explosion is the impact that occurred when the lifted shield plug landed.

IV. The fire engine pump injected not that much water, and it did so slowly. As a result, the reaction between the mixed molten substance and water was not active, but rather occurred slowly. Even under such a condition, there is no evidence that the core melted down to flow out of the reactor pressure vessel. During this time, the hydrogen that leaked outside of the reactor building and/

or the hydrogen gas discharged through the vent flowed backward into the reactor building of Unit 4 via the exhaust facility to become the cause of the explosion of Unit 4.

V. It seems that a core melt would not have occurred in Unit 3 if HPCI was not stopped (hence maintaining the pressure at a lower level and allowing the water injection to be made easily), and the water injection was conducted accordingly.

2.7 Back to the Case of Unit 1

The reason I interrupted the description of Unit 1 earlier was that it was not clear to me what happened to the core melt condition at the time when the core lost water completely. That is why I decided to study the melting of Units 2 and 3 first and then make a deduction from those analyses. Let us review those cases here.

After losing the external power source, Unit 1 was switched to manual control cooling by means of IC, but the unit was hit by the tsunami immediately after the isolation valve was closed. The emergency power was made unusable because of the tsunami, thus leaving IC turned off. If they noticed that the IC had stopped soon enough and taken countermeasures, the history of the Fukushima accident could have been substantially different. It was an unfortunate missed opportunity.

In the meanwhile, the reactor that lost its means of cooling was using water evaporation for removing decay heat, as if a hungry octopus eats its own tentacles. The reactor's cooling was maintained by opening and closing the safety relief valve to release the generated steam. However, this octopus-tentacle-eating measure cannot last long. This has to end when all the water inside the reactor is lost.

With the evaporation, the reactor water level dropped with time. According to a calculation by TEPCO, the water level that was plus 5 m above the core in the beginning lowered to the top of the core (water level index: 0) about 3 h later (about 6 p.m.), and then to the lowest part of the core about 5 h later (about 8 p.m.). With some corrections, this calculation can be trusted. By midnight, the core lost water completely.

2.7.1 Radiation Heat in Reactor Pressure Vessel

When all the water in the core is lost, heat from the core will be released through heat radiation from the core, as opposed to heat transfer from the fuel rods to water. Just like the sun evaporates water, the heat radiation from the red-hot core evaporates the water below the core. However, since the radiated heat is not only absorbed by water, but also reflected, so not all of the heat gets transferred to water as in the case of thermal conduction. Moreover, the radiation heat irradiates everything in the vicinity, not just water. Thus, the reactor's heat dissipation changes the scheme entirely.

If we look at this from the standpoint of the fuel rods, the loss of water means that the heat removal totally depends on the radiation heat. Since the radiation heat emits in proportion to the fourth power of the absolute temperature, the heat source's own temperature needs to be raised in order to dissipate heat efficiently. Thus, the fuel rod temperature becomes much higher than when it was immersed in water.

Furthermore, as adjacent fuel rods radiate heat with each other, their exchange of heat is a net zero in the end, so that the entire set of fuel rods (core) becomes a one-temperature (although there may be slight deviations) group that radiates heat to the internal components.

More specifically, the temperature of the core (fuel rods) that has lost water rises sharply as it dissipates the decay heat by radiation heat. The temperatures of the internal components rise gradually as they receive said radiation heat, thus causing them to dissipate heat themselves, resulting in a chain-reaction, heat dissipation system. This heat dissipation system became stabilized after a fair bit of time passed from when the reactor pressure vessel completely lost water.

As the core loses its water and the heat dissipation system starts to change from a conduction system to a radiation system, the heat that is used for evaporation of water should be reduced by the amount that is used for heating the core and the internal components, so that the rate of water level drop slows down as well and the timing of the complete loss of water from the core will be delayed. I believe that this delay should be around 1–2 h.

According to the TEPCO's calculation, the reactor water level is estimated to have reached the very bottom of the reactor pressure vessel, i.e., minus 8 m, at around 11 p.m. on March 11, but I estimate rather that it was around midnight when the reactor pressure vessel lost its water completely.

That is the review of Sect. 2.4 of this chapter.

In case of the TMI accident, a lot of cooling water was injected when the core water level was only halfway down. Consequently, the core melt occurred immediately because of the zirconium-water reaction. In case of Unit 2, the core collapse and the core melt occurred as a result of the zirconium-water reaction when the cold seawater was injected after the reactor water level had dropped to as low as 1 m below the bottom of the core due to decompression boiling. The core melt of Unit 3 occurred after the core lost most of its water, the reactor pressure was reduced, and the oozy zirconium-water reaction had been occurring for about 24 h.

Common to all three reactors is the fact that the cause of the melting is the zirconium-water reaction, and that the core melt occurred when there was water in the reactor. However, as described before, the water level was halfway down the core in case of TMI, was near the very bottom of the core in case of Unit 3, and was 1 m below the core in case of Unit 2. This difference is very important in thinking about the heat dissipation of the core. In case of TMI, because there was water in the core, the main way in which the core was cooled was transferring heat to water, whereas there was no water left in the core of Unit 2, and so the core melt occurred while heat was being dissipated through heat radiation. The intermediate case was the core melt of Unit 3. In both Units 2 and 3, the core had lost its water and the heat was dissipated by heat radiation. In the case of Unit 2, this condition did not last long, and its core melted although the core essentially retained its shape. In case of Unit 3, oozy reactions continued for more than 24 h in the core, which was partially collapsed. However, in any of those cases, there was no difference in that the oxidation reaction occurred only when the high temperature fuel rods (core) met a large amount of cold water and the reaction heat caused the core to melt. The above discussion is the starting point of the analysis of Unit 1. In Unit 1, in which the reactor pressure vessel had lost its water completely, the heat radiation is the only means of removing the decay heat generated in the core. It is different from Units 2 and 3 in that sense.

Let us examine now what happens inside a core which has lost its water.

The decay heat was the heat generated as a result of radiation attenuation. There are three kinds of radiation, α, β and γ-rays. As α- and β-rays do not penetrate materials, heating occurs on the spot. However, γ-rays penetrate substances, so that heating occurs not only in the reactor but also in the pressure vessel walls and shielding concrete.

Since the fuel rods (core) are surrounded by water when the reactor is operating, γ-rays attenuate (generates heat) in the water and only a small portion of the γ-rays penetrates through to the reactor pressure vessel. Therefore, the α, β and γ-rays are treated together as a group as the source of the decay heat in the reactor. However, the situation is different in a core which has lost its water. Although a portion of the γ-rays is attenuated in the structural members surrounding the core, contributing to their heating, the majority of the γ-rays is attenuated (causes heat) in the reactor pressure vessel and the shielding concrete. Therefore, the heating by γ-rays, which amounts to one half of the total decay heat, needs to be considered separately from the heating of the core.

More precisely speaking, even after the water is gone, so long as the core maintains its shape much of the γ-rays attenuate in the process of penetrating through the surrounding fuel rods, and so the heating inside the core does not decrease so much. However, once the core loses its original shape, more γ-rays radiate outside so that the heating inside the core reduces. In the end, when the core loses its original shape and reduces to a lump of molten core, putting aside dissipation within the lump it is reasonable to think that virtually all of the γ-rays radiate outside. If that occurs, I believe as a ballpark estimate that about 70–80 % of the heating by γ-rays occur outside of the core.

If we look at this from the standpoint of the decay heat amount generated inside the core, since γ-rays account for half of that, and about 70–80 % of the γ-rays heating occurs outside of the core, the amount of generated heat is reduced by 35–40 %. This means that the decay heat that heats the core is reduced to two-thirds of the original amount.

In other words, an interesting phenomenon occurs when the core changes to the radiation-dominant phase, in that the decay heat that heats the core is reduced to two-thirds. Of course this change does not occur all at once, but rather occurs gradually caused by the combination of the lowering of the water level and the collapse of the core shape, and I will use the term "radiation heat phase," in this book to refer to the phase after the reactor water level has reached the core bottom as radiation will become the main mode of heat dissipation in that phase.

The core lost its cooling water at around 8 p.m. on March 11. Unit 1 moved into the phase of heat removal by radiation. Once it loses its means of cooling, the core temperature starts to rise rapidly and emits the radiation heat. Together with that, the temperatures of the internal components start to rise as well. What kind of effects did this change have on the reactor?

According to the TEPCO report, the reactor pressure, which used to be approximately 7 MPa at around 8 p.m. on March 11, dropped to 0.9 MPa by 2:45 a.m. on March 12. Those are the only two points in time when the reactor pressure data of Unit 1 were recorded. It is unfortunate that we don't know what happened in between, but it is at least certain that it is one change that the radiation heat phase caused.

The only Unit 1 data available to us is the containment vessel pressure shown in Fig. 2.19. And even this only shows the condition after midnight (around 1 a.m.). The reactor water level data is not reliable either. Given the lack of data, the rest of my descriptions include a substantial amount of speculations.

It can be presumed based on the case of Unit 2 that the original shape of the core was maintained until around 8 p.m. on March 11, or just before the core ran out of water, with the fuel rods cooled by the small amounts of steam that were flowing. It is confirmed that the reactor pressure was 6.9 MPa at this time because the operating staff read the instrument directly at site.

From this we can deduce that the core was intact until 8 p.m. on March 11.

The decay heat at the time when the core lost water (around 8 p.m.), by the way, was about less than 1 % of the rated power, as it was approximately 5.5 h after the

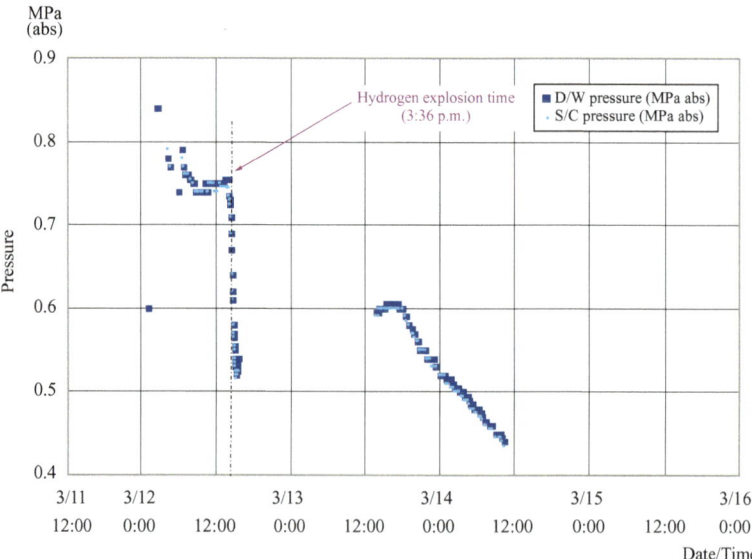

Fig. 2.19 Unit 1 containment vessel pressure change (measured data) (Source: from "TEPCO Nuclear Accident Investigation Report")

reactor stopped. Since the radiation heat is about two-thirds of that, that means approximately 0.7 %. As the rated thermal output of Unit 1 is approximately 1,500 MW, 0.7 % of that is approximately 10 MW. This amount of heat is capable of evaporating in 1 h approximately 25 tons of cooling water at the rated temperature and pressure, or approximately 13 tons of cold seawater. As the size of the core is approximately 3 m in diameter and approximately 4 m in height, the volume is about 30 m^3. For this small, water-less space, 10 MW of generated heat is an enormous size. These numerical values will be used later.

It was reported at 9:51 p.m. that the radiation level of the reactor building was increasing.

It was already 2 h since the water ran out in the core. This rise of radiation means both that radioactive substances had started to leak out from the fuel rods and also that such radioactive substances had leaked into the containment vessel. Both of these were things that we had not experienced before. These are probably the result of the reactor having entered the radiation heat phase. Let us carefully examine the status inside the core.

I suppose that the fuel rods in the core center, where the heat generation was most active, had started to deform. The fuel rods must have started to deform, bending and buckling, contacting and leaning with each other, and started to form a mixed molten substance consisting of uranium dioxide and zirconium, drawing into it the surrounding channel box as well.

I know that I have not yet mentioned the channel box. The channel box comprises rectangular outer plates that surround the fuel assembly of BWR, each consisting of a zircaloy plate having a width of approximately 14 cm on each side and a thickness of 2–3 mm (Fig. 2.20). As water and steam flow inside the channel box, when there is water in the core, the temperature of the channel box, which does not generate heat by itself, is low, so I have been treating this as something that is different from the fuel rods.

However, once the reactor is in the radiation heat phase, the temperature of the channel box which is made of thin plates becomes equal to the temperature of the fuel rods within a relatively short period of time.

Fig. 2.20 Structure of fuel assembly (Source: from "Outline of BWR Power Plant" (Rev. 3) by NSRA)

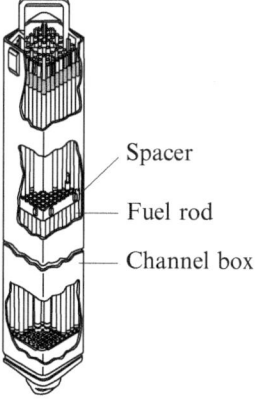

Spacer

Fuel rod

Channel box

Allow me to digress a little, but the amount of zircaloy used in the channel box is too large to ignore. Although it varies with the type of fuel rods, the proportion of the zircaloy in the core that is attributable to the channel box is 36 % where the channel box is 2 mm thick and 45 % where it is 3 mm thick.

As described later, the fuel rod temperature during the radiation heat phase is estimated to have reached around 2,000 °C by 4 a.m. of the following day. To our chagrin, this temperature is slightly in excess of the zircaloy melting point, so in the radiation heat phase the large amount of zircaloy in the channel box will be part of the core melt.

By this time, the entire surface of the channel box must have been covered by zirconium oxide film as a result of the reaction between zirconium and steam. In the meantime, however, the main part of the zircaloy sandwiched between the oxide films melts at around 1,800 °C and drops down by gravity. This process cannot be taken lightly. It seems that the temporal temperature change of the channel box significantly affects the core melt and the hydrogen explosion in the radiation heat phase.

The internal components in the vicinity of, and above the core are mostly made of a stainless steel alloy that has a low melting point (1,450 °C), so that they must have melted partially by the strong radiation heat from the core and flowed down on the mixed molten substance little by little. Thus, as the internal components in the vicinity of the core melt down, the radiation by the mixed molten core starts to heat the reactor pressure vessel directly.

On the other hand, the evaporation of water by the radiation heat must have been continued constantly and the water level presumably dropped gradually. However, the lower internal components of the core are thought not to have melted as they were kept relatively cool because their lower parts were still immersed in water.

That is the status of the reactor at around 10 p.m.

An hour later, at around 11 p.m., it was recorded that the radiation level was high in front of the double door in the reactor building. This is evidence that a substantial amount of radioactive materials leaked out into the containment vessel. According to the TEPCO calculation, the water in the reactor pressure vessel was completely depleted by this time.

It was confirmed that the DW pressure of the containment vessel reached 0.6 MPa at around 11:50 p.m. That means that an opening was formed in the reactor pressure vessel due to the heat, and the steam inside leaked out from the reactor.

Let's see if it is true by making a simple calculation. The time, 11:50 p.m., is about 4 h since 8 p.m. when the water level reached the very bottom of the core. If we assume that all of the radiation heat was used to evaporate water, it can be concluded that approximately 100 tons of water was evaporated within these 4 h since the evaporation amount is 25 tons/h according to the calculation we did a while ago. This evaporation amount is slightly more than the quantity of water retained below the core. I suppose that the reactor water was completely depleted by this time, i.e., midnight. The opening in the reactor pressure vessel was probably formed approximately this time.

That was the status of the reactor in the radiation heat phase, at around midnight on March 11.

2.7.2 Core Melt and Meltdown

What did the opening in the reactor look like?

I think the opening here is a gap that was formed when the top lid of the reactor pressure vessel was lifted up due to internal pressure when the fastening bolts used to keep it in place got elongated by thermal expansion on account of the top portion of the pressure vessel being directly or indirectly heated by the radiation heat. I surmise that steam leaked through that gap into the containment vessel. To put it simply, the reactor pressure vessel's top lid got loosened by thermal expansion and a gap was formed.

Some people imagine that the safety relief valve got hot and the valve closing force weakened. That is also a viable opinion.

There are various other theories, including a theory that says the flange coupling of the main steam piping that extends from the top of the reactor pressure vessel was damaged by high temperatures, a theory that the control rod-drive housing located at the bottom of the core or the neutron measuring piping was damaged by melting due to the drop of the molten core, etc. All of them are credible. I think you also should use your imagination freely. I believe that there is no question that an opening was form in the reactor pressure vessel by the time the new day started and the pressure vessel became connected to the containment vessel to allow the gas to leak to the latter.

The condition of the core must have been such that the mixed molten substance, which included molten upper components, had developed here and there during from 8 p.m. to midnight, and the entire core must have reached a very high temperature, although it had not completely collapsed. I have no choice but to think so since it had lost its coolant water. During this period, a portion of the nearly 100 tons of water that evaporated must have caused a oozy oxide reaction with the mixed molten substance and developed some hydrogen gas as well. Since this hydrogen gas is light and hot, it must have gathered around at the top portion of the reactor pressure vessel and heated its vicinity. Because of this core condition, the opening in question must be the loosening of the top lid of the reactor vessel.

This was the state of the core from between 8 p.m. on March 11 and 0 a.m. on the following day.

Let us now move the hands on the clock and think about the reactor's condition during the period from 12 a.m. through 4 a.m. on March 12. Since all the water in the reactor pressure vessel had already evaporated, the temperature of the mixed molten substance made from the core must have risen further, getting closer to its melting point. However, the core as a whole at this hour was probably on the core plate, although it had deformed substantially. My thought is based on the result of the PCM-1 test.

Because of the thermal effect of the radiation heat of the core and the molten upper core materials, probably the guide tubes of the control rod drive mechanism and the neutron measuring device were partially melted and holes developed at the bottom of the reactor pressure vessel. Although the top lid of the reactor pressure vessel still remained, its pressure keeping capability was completely lost.

The condition of the containment vessel was such that the temperature of the top portion of it had probably substantially increased from the superheated steam and hydrogen gas from the reactor. However, the containment vessel is such a huge structure so that the temperature rise occurred only in the top portion. It is difficult to believe that the entire structure was uniformly heated. Moreover, the radiation heat emanating from the reactor pressure vessel towards the containment vessel was shielded by the insulating material covering the reactor pressure vessel, so I think that it was not contributing much to the heating.

I believe that, in general, a serial radiation heating system was about to be formed consisting of the core which turned into the mixed molten substance, the reactor pressure vessel, the insulation material, and the containment vessel.

From 0 a.m. to 4 a.m. on March 12, when the water injection using the fire engine started, there was a span of 4 h. Yet, we have no data for that period. No decisive clue. The only thing we know is the physical fact that the decay heat dropped just a little bit.

There are many people who proclaim theories on how the mixed molten substance behave during this 4 h period, in particular that the substance partially or fully melted the bottom of the reactor pressure vessel and dropped in a liquefied state on the floor of the containment vessel. Those claims sound convincing. However, nobody seems to be able to tell us about the physical condition, e.g., how much material dropped, did it only drop once, or rather did it drop continuously, was it fluid, or was it in a squashy state just before melting, etc. It seems that they only claim that it melted and dropped somehow without any convincing reasoning, and I cannot blame them too much for that. It is because that there is no data on which to build presumptions.

2.7.3 Reverse Study from Hydrogen Explosion

I suppose that the proper thing to do in this kind of situation is to think backward. Let us try to think backward from the explosion, which is the final moment, leaving aside the phenomena between 0 a.m. to 4 a.m. on March 12 for the time being.

The explosion occurred in the reactor building of Unit 1, at 3:36 p.m. on March 12. The only thing that was damaged by this explosion was the fifth floor where the fuel exchange floor is located, exactly above the reactor. This is the unique feature of the explosion of Unit 1. The explosion in Unit 4 damaged not only the fifth floor but also the third and fourth floors, while even the 1st floor damage is observed in case of the explosion in Unit 3.

The explosion of Unit 1 affected only the 5th floor. This is the first clue of the backward analysis.

Figure 2.3 is the cross-sectional drawing of the reactor building. As we look at it closely, we see that the fuel exchange floor is directly above the containment vessel.

The fact that only the fuel exchange floor exploded means that the hydrogen gas existed only on the 5th floor. The only route that makes the hydrogen gas flow only to the 5th floor is the one that makes the hydrogen gas flow directly upward from the containment vessel.

The only hydrogen path available is, as I described in the explosion of Unit 2, for the hydrogen gas from the containment vessel to raise the reactor vault pressure, push up the shield plug, and flow into the fuel exchange floor. If the hydrogen gas had passed through other routes, the hydrogen gas left in those routes would have caused chain explosions. The flow of the hydrogen gas that lifted the shield plug, which I explained in connection with Unit 2, occurred in Unit 1.

Or rather, the fact that the explosion occurred only on the 5th floor in Unit 1 showed us how the hydrogen flowed from the containment vessel to the reactor vault and then to the fuel exchange floor. I have to admit that even my reasoning here was backward because the order in which I wrote this was backward.

In the case of Unit 1, there are many people who believe that the hydrogen gas first entered the stack from the vent as the vent was opened about 1 h prior to the explosion, and then backed up via the ventilation duct. This is wrong. In the case of Unit 4, the explosion was caused by the hydrogen back-flow from Unit 3, but it had a long time – as long as 26 h – to back-flow. In case of Unit 1, the time the vent was open before the explosion was only 1 h. It is unreasonable to claim that a sufficient amount of gas for causing an explosion can flow backward from the stack to the fifth floor in such a short time. Also, the explosion in case of Unit 1 occurred only on the 5th floor, different from Unit 4. There was no explosion on the lower floor where the ventilation duct is located.

That makes it necessary for a large amount of hydrogen gas to flow into the reactor vault in order to push up the shield plug before the explosion occurs. The containment vessel pressure was as much as 0.5 MPa even when the vent was open at around 2:30 p.m. on March 12, which is large enough relative to 0.05 MPa, the pressure required to lift the shield plug. The remaining mysteries are why and when the top lid of the containment vessel opened.

There is no question about the timing. It was immediately before 3:36 p.m., when the explosion occurred. I will explain why it opened later.

Without an ignition source, there can be no explosion. Since the power station is under a power outage, the only ignition source can be the vibration from an earthquake or the shock from something dropping. TEPCO confirmed that there was no earthquake at that time. The only possible ignition source is the shock generated by the drop of the shield plug, as in the case of Unit 3.

That means that the time that the explosion occurred, i.e., 3:36 p.m., is when the shield plug dropped. It means that the top lid of the containment vessel opened just before then, causing a large amount of hydrogen gas to enter the reactor vault to lift the shield plug. The top lid of the containment vessel must have opened widely just prior to that time. The only cause that could open the top lid of the containment vessel was the generation of hydrogen gas due to the oxidation reaction between the mixed molten substance and water.

However, for some unknown reason, the containment vessel pressure data (Fig. 2.19) that could have provided the evidence for such events is lost for a period starting around 1 h before the explosion. The route to prove facts based on backward analysis is also closed due to a lack of evidence. I can only proceed with my analysis from this point on by mixing in presumptions. Please keep that in your mind.

First, I will try to estimate the temperature under the radiation heat phase during the period of 4 h from 0 a.m. to 4 a.m. on March 12. The only thing that remains in the empty reactor pressure vessel now void of all water is the radiation heat. The strength of the radiation heat is a little less than 10 MW and the source of the heat is a cylindrical object of approximately 3 m in diameter and approximately 4 m in length, a mixed molten substance which used to be essentially the core, comprising uranium, zirconium, oxygen, stainless steel, etc. Although it is difficult to estimate, let us move forward with our calculation assuming that the core temperature at around 4 a.m. on March 12 was 2,000 °C, which is close to the melting point of the mixed molten substance based on the example of the TMI accident.

Thus the reactor which lost cooling water is generating 10 MW of radiation heat. As I mentioned previously, an equilibrium status of the radiation heat, or a serial radiation heating system consisting of the core, the reactor pressure vessel, and the containment vessel, was formed. Let us do some calculations about the temperature of this system.

The calculation, a very rough one I admit, is shown at the end of this chapter (Appendix 2.8). If the core temperature is assumed to be 2,000 °C, the equilibrium temperatures achieved by the radiation heat are approximately 550–600 °C as the reactor pressure vessel temperature and 120–130 °C as the containment vessel temperature. From this calculation result, we can be sure that the temperatures of the reactor pressure vessel and the containment vessel will not be unrealistically high even if the core temperature is as high as 2,000 °C.

If I may add, if there is a 10 % error in the assumption of the core temperature, the error in the radiation heat will be as much as 50 %. Using the fact that radiation heat is proportional to the 4th power of temperature as a hint, please confirm the above.

Then the next question is how long it would take to reach an equilibrium state of heat radiation. Let us check that time as well.

In terms of assumptions for our calculation, let us assume that the reactor pressure vessel, which weighs about 340 tons, has reached 550 °C, one half of the reactor internal components (the portion from the core up), which is estimated to weigh about 100 tons, melted by the radiation heat, and the rest of the internal components were heated to 550 °C – as they can dissipate heat because components such as the control rod drive mechanisms extend to the outside of the reactor pressure vessel (Appendix 2.9). Since the outside of the reactor pressure vessel is covered by a thin aluminum insulation material (melting point: approximately 650 °C), the heat radiation from the reactor pressure vessel to the outside is assumed to be zero in the calculation.

The above calculation indicates that a heat quantity of approximately 30 MWh will be needed to reach the equilibrium state. The core radiation heat at around

0 a.m. on March 12 was approximately 10 MW. Since the core radiation heat at around 4 a.m. on March 12 was 7 MW, the total radiation heat generated during these 4 h was slightly less than 40 MWh. If we compare both, the radiation heat appears slightly larger but in essence both are roughly equal considering heat dissipation, core temperature increase, etc.

This means that by 4 a.m. on March 12, when the water injection was started, the radiation heat equilibrium temperature system consisting of a core temperature of approximately 2,000 °C, a reactor pressure vessel temperature of 550–600 °C, and a containment vessel temperature of 120–130 °C was more or less established. The core temperature of 2,000 °C seems to have been appropriate judging from the calculation result. We can assume that the core has not yet started melting at this time.

2.7.4   Drop of Partially Molten Core and Its Internals

Once the temperature estimation is made, next is the estimation of the internal state of the core.

The melting point of the core fuel, i.e., uranium dioxide is 2,880 °C. That of stainless steel is approximately 1,500 °C. The two are vastly apart. The melting point of the mixed molten substance comprising uranium, zirconium and oxygen is approximately 2,000–2,200 °C. As you can see, the core toward the equilibrium temperature system comprised materials with widely varying melting points.

The matter would have been simpler if these materials were melting separately. Under the 2,000 °C core temperature state, stainless steel would have melted by the radiation heat and flowed down like water to the bottom of the reactor pressure vessel. The uranium dioxide would have remained as solid pellets. The mixed molten substance would have turned into a soft plastic object immediately before melting.

However, as the existence of the mixed molten substance consisting of uranium, zirconium and oxygen suggests, high temperature metals mix before melting and co-melt with each other, so that the result is quite unpredictable. Even stainless steel that can melt and flow down like water may react when it contacts a substance just before it melts and may produce an alloy we are not aware of. For a person like myself who has been brought up in a world of physics that can be dealt with using mathematical formulas, the reactions of high temperature metals joining and separating with each other seem, sorry to say, very mysterious and outlandish. To me they are very unpredictable.

Moreover, these mixed molten substances and other mysterious substances are sitting on a thick perforated stainless steel plate called the core plate. Since it is perforated, it allows fluid materials to pass through. It is presumed that molten stainless steel and such passed through the perforated holes and dropped to the bottom of the reactor pressure vessel and accumulated there. The molten stainless steel must have cooled at that bottom head of the reactor pressure vessel and solidified again. At the same time, the bottom head of the reactor pressure vessel must have been heated and softened because of the heat.

On the other hand, solids such as uranium dioxide and softened plastic components such as the mixed molten substance do not drop easily as they are supported
by the thick core plate. Depending on their fates, those highly heated and plasticized
substances may drop below as lumps, be mixed and melt to form alloys we have
never seen, or survive as solids. One of the more probable scenarios is that a portion
of the lower internal components is melted by heat and drops to the bottom of the
reactor pressure vessel mixed with some of those indescribable formations. It is
possible that the bottom of the reactor pressure vessel, which had lost its strength
because of heating, broke down from the weight of those lumps dropping on it and
then all of that dropped to the floor of the containment vessel floor.

Or rather, its drop could have been interrupted by obstacles such as the control
rod device mechanism and the like, or melted the obstacles and piled on the bottom
of the reactor pressure vessel. It seems to me that innumerable varieties of events
could have occurred inside the reactor pressure vessel – the center of this mess,
so-to-speak, during this period of time.

The pipes that go through the bottom of the reactor pressure vessel include the
penetration part of the neutron measuring system as well as the penetration part of
the control rod drive mechanism. Many people have pointed out the possibilities
that various kinds of piping melted or broke down and that produced small cracks
and holes at the bottom of the reactor pressure vessel.

I imagine that such conditions started around 4 a.m., or just before the water
injection was started, although the precise timing may have been a bit different. As
seawater is injected by the fire engine pump, an oxidation reaction starts between
the highly heated mixed molten substance and the cool seawater. Since the oxidation reaction generates heat, the core melt accelerates automatically. The problem is
whether the water injection was sufficient.

The water injection by the fire engine started at 4 a.m. on March 12. The TEPCO
report notes that approximately 80 m^3 of water was injected by 2:53 p.m., just before
the explosion of Unit 1. This amounts to 80 tons in about 11 h, or only 7 tons/h, but
based on what follows let us assume a rate of 5 tons/h. Since it had been almost a
half day since the reactor stopped by this time, the decay heat had reduced to about
0.7 % (approximately 0.7 MW in terms of the decay heat inside the core). Since this
amount of heat can evaporate approximately 10 tons or more of cool seawater,
the seawater injection by the fire engine of 5 tons/h does not even come close in
matching the decay heat. That means the core temperature keeps rising. The mess in
the reactor pressure vessel I described above must have been further intensified by
this time.

It can be generally said of pumps that a higher discharge pressure results in a
smaller amount of discharge. For 10 h from 4 a.m., when the water injection started,
until 2 p.m., the containment vessel pressure was approximately 0.8 MPa, which
was about equal to the pump discharge pressure. That is why the pump discharge
was so small. However, as the SC vent was opened at around 2:30 p.m., the containment vessel pressure dropped sharply to about 0.5 MPa (Fig. 2.19). When the discharge pressure drops to 0.5 MPa, the pump discharge amount increases sharply. In
order to further the discussion, let us assume approximately 30 tons of seawater,

which is approximately one-third of the total 80 m³ seawater injected, entered the reactor pressure vessel in such a state (Appendix 2.10). This means that the rate of water injection during the foregoing 10 h was 5 tons/h.

Incidentally, it is reported that the spray of seawater in Unit 1 was executed using the core spray which sprays the core with water from above, but the stainless steel piping on top of the reactor is considered to have been melted by that time, so that the water must have run down along the core's external wall surface. It is estimated that the majority of such a small amount of seawater, or 5 tons/h, was evaporated by the surrounding heat and some of the steam thus generated must have reacted with the high temperature mixed molten substance and the heat provided by said reaction must have further elevated the core temperature. I suspect that the dropping of the molten lumps mentioned before must have been quite active as well. All in all, however, the reaction itself was not that active because of the insufficient availability of water. We can consider this as a high-temperature version of the oozy reaction mentioned in the description of Unit 3.

If we are to analogize this semi-molten core status, it was like well-cooked and softened chestnuts embedded in the mushed sweet potato of a traditional Japanese sweet served on New Year's Day. The majority of the uranium dioxide pellets were soft but still solids. On the other hand, the mixed molten substance had become plastic, just like the main body of the Japanese sweet. This Japanese sweet in our analogy is protected by the thin incomplete egg-shell made as a result of the zirconium-water vapor reaction from the direct contact with water. I suppose that is how it was lying on top of the core plate. I wonder if this was when the bottom of the reactor pressure vessel fell through.

Although all kinds of conjectures are possible, my thinking is as follows: "At a certain point in time, a portion of the mixed molten substance, which had a certain amount of weight, fell through the bottom of the reactor pressure vessel and dropped on the floor of the containment vessel. The surface of what fell got cooled by the atmosphere of the cooler containment vessel floor and solidified. On the other hand, the remaining mixed molten substance left in the core reacted with steam vapor, widened the egg-shell, and remained on the remaining part of the perforated plate. I suppose, however, the heated and softened mixed molten substance still kept dropping in clumps from the cracks of the incompletely formed egg-shell, eventually forming a mound."

To reveal my secret about this discussion, it is at least partially based on the deduction from the Chernobyl core melt case. As it is described in Sect. 1.3.1, the fuel left after the core fire in the Chernobyl accident is considered to have melted the bottom of the concrete reactor vessel, dropped to another floor 2 m below, and produced a mound. Further melting and fluidization by the decay heat occurred inside of the mound, and the fluidized substances broke out the edge of the mound and flowed out like a flood three times.

I suppose a similar situation continued for 10 h. The mixed molten substance that dropped on the containment vessel floor must have been cooled by the cool atmosphere of the floor so that the outside walls must have solidified. However, the temperature in the inside must have risen further due to the continuing decay heat, and the mixed molten substance must have gradually melted. I am sure that the situation of the lower part of the containment vessel at that moment was quite confusing, with various ingredients partially and completely melted and mixed together.

2.7.5   Sea Water Injection and Explosion

At around 2:30 p.m. on March 12, the vent finally opened, and the pressure of the containment vessel dropped sharply. The water from the fire engine kept running. All the work-related people who worked all through the night must have gotten quite relieved. However, as soon as they were relieved, another monster appeared. It was an explosion that occurred at 3:36 p.m.

The explosion was caused by the large amount of hydrogen generated by the active reaction between cold water and the extremely hot mixed molten substance. The active reaction took place due to the increased injection of seawater. The rate of seawater injection which used to be 5 tons/h became 30 tons/h in a short time. The seawater, which had up to then evaporated, started to fall on the floor of the containment vessel floor as liquid water.

The seawater ran down along the core's external walls. Although there is no record to prove it, I assume that a large amount of hydrogen must have been produced by the reaction between water and the potpourri of molten alloys in the mound.

This assumption is basically correct but there is one thing that does not match the data. It is a time difference of about 1 h between the time the water injection amount was increased (a little past 2:30 p.m.) and the time of explosion (3:36 p.m.). The presumption that can solve this difference is the deduction from the Chernobyl accident, the mound on the floor. There is only one clue here: the fact that a large amount of water is needed at once in order to cause a violent reaction. Please remember the core melt in TMI as well as the Units 2 and 3 accidents.

The majority of the water that dropped on the containment vessel floor most likely must have dropped on the outside of the mound, and must have been unable to react immediately with the molten alloys as they were kept separate by the mound. The lower part of the reactor pressure vessel has many things, such as the control rod drive mechanism and neutron measuring cables, that must have been hanging down like drainpipes, so that the dropping water must have flown along them to the outside of the mound. They must have followed a route that was different from the straight-line route that the dropping semi-solid lumps followed.

The violent reaction started around 3:30 p.m., when the water level on the floor rose to exceed the height of the mound. The delay of about an hour from when the vent started must have been the time taken for the water level to rise. This is a situation similar to the Unit 3 reaction that happened three times. I wish I could write that's the case but I have no data to prove it.

The reaction between the high temperature mixed molten substance and a certain amount of seawater must have caused the containment vessel pressure to rise sharply (TEPCO's record is cut short). This pressure rise lifted the top lid of the containment vessel. The preheated fastening bolts of the top lid must have been elongated by thermal expansion and then further elongated by the lifting of the top lid. The pressure may have temporarily risen to more than 0.8 MPa.

Once we are convinced of the lifting of the top lid of the containment vessel, I suppose that there is no need to repeat the description of the process of the hydrogen entering the fuel exchange floor of the 5th floor of the reactor building, mixing with air to become an explosive gas, and igniting by the impact of the dropping of the shield plug.

So that is the presumption of the core melt and explosion in Unit 1.

The uniqueness of the explosion of Unit 1 is that it was limited to the 5th floor and the scale of the explosion was smaller than that of Unit 3.

The reason that the scale of the explosion was smaller was that the amount of hydrogen gas that contributed to the explosion was relatively smaller. This is believed to be caused by the fact that both the amounts of the mixed molten substance and water were smaller. Furthermore, a portion of the hydrogen gas that was generated on the containment vessel floor was discharged to the atmosphere through the vent as the vent of Unit 1 was open at the time of the explosion.

The water injection continued even after the explosion. There is no question that the reaction between the mixed molten substance left in the core after the explosion and steam continued. The reaction must have continued even after the core was flooded with water. However, the hydrogen gas thus generated after the explosion was diluted by a large amount of atmosphere, as there was nothing left to cover the containment vessel because of the explosion, and did not become an explosive gas.

In the case of Unit 3 explosion, immediately after destroying the 5th floor of the reactor building in horizontal directions, split the building vertically from around the 3rd floor. Its blast is said to have reached several hundred meters high into the sky. On the contrary, the explosion of Unit 1 occurred only on the 5th floor.

Actually, the fact that the explosion of Unit 1 occurred only on the 5th floor provided us the evidence for identifying the discharge route of the hydrogen gas. In other words, the route by which the hydrogen gas entered the reactor vault by pushing up the top lid of the containment vessel, lifting the shield plug by its pressure, and then going out onto the fuel exchange floor on the 5th floor was proven originally by the analysis of Unit 1. If the hydrogen gas went through a different route, the explosion of Unit 1 must have caused an explosion along the route, just like what happened in Units 3 and 4. The mysteries of the hydrogen gas flow routes and the causes of ignitions of the hydrogen gas explosions in Units 2 and 3 were solved with the hint provided by Unit 1 – although the explanation was provided in reverse order.

Although the explosive power was weak, the explosion of Unit 1 significantly affected the accident countermeasures of the other reactors. The overnight effort of directly providing electric power to Unit 2 was nullified by the damages caused by the explosion to the power source car and the connecting cable. Then, Unit 2 developed the core melt because the RCIC pump stopped 2 days later around 11 a.m. on March 14.

The explosion of Unit 1 did not cause any further damage to Unit 1, but it kicked off a major calamity to the entire Fukushima Daiichi Nuclear Power Plant.

2.7.6 Section Conclusion

The following is the summary of the important conclusions concerning the melting and explosion of Unit 1.

Unit 1 stopped as the earthquake occurred and was in the IC cooling state by manual operation, but unfortunately it was hit by a tsunami when its cooling was stopped so that it became impossible to restart it. It was further unfortunate that this information of cooling failure was not communicated to the supervising staff in a proper way, so that both the site manager and the consulting staff at TEPCO's head office misunderstood that Unit 1 was OK because the IC was working. As a result, they went into the accident without countermeasures. That was the biggest cause that invited the core melt and the explosion in Unit 1.

After the tsunami struck, the Unit 1 reactor had to keep cooling the decay heat without a drop of water supply. The only thing it could do was to evaporate the cooling water left in the reactor pressure vessel, akin to a hungry octopus eating its own tentacles. The water level of the reactor kept dropping with time, and it became empty by midnight on March 11. What happened in Unit 1 has never occurred in any other nuclear power plant. Even under such a reactor condition, the core has not been melted. This is a remarkable fact.

The only means of heat dissipation available once the reactor pressure vessel loses its water completely is to rely on heat radiation. The higher the temperature, the more heat is radiated. In order to be able to succeed in heat removal, the core temperature of the heat source needs to be very high.

It is only a rough estimate, but the core temperature of Unit 1 rose to 2,000 °C, and the reactor pressure vessel temperature rose to about 550–600 °C. In the meanwhile, a portion of the internal components of the core made of stainless steel must have melted, and the lower core support structure must have deformed as well. I suppose that various conduits that are passing through the bottom of the reactor pressure vessel head such as the guide tubes of the control rod drive mechanism and the neutron measuring system were melted as they met with the down flow of molten stainless steel, thus forming the openings for water leakage.

Around 4 a.m. on March 12, the water injection by the fire engine started. However, the amount of water injected was small because the pressure of the reactor pressure vessel was too high. Moreover, most of the injected water evaporated in the reactor pressure vessel. This steam and a portion of the high temperature core caused the oozy reaction and continued to generate heat and hydrogen gas for about 10 h.

The generated hydrogen gas accumulated on the top portion of the containment vessel, and heated the top lid of the containment vessel. Due to the temperature rise, the fastening bolts of the top lid thermally expanded and weakened the fastening force.

In the meantime, the heat from the oozy reaction in combination with the decay heat rose the temperature of the mixed molten substance and caused partial melting. A portion of the molten substance is assumed to drop to the bottom of the reactor pressure vessel together with the molten core plate. The bottom of the reactor pressure vessel which had been heated to a high temperature by the radiation heat and the conductive heat of the molten stainless steel became unable to withstand the weight of what had dropped, and dropped to the floor of the containment vessel together with the other molten droppings.

Once a hole is opened in the bottom of the pressure vessel, the already softened mixed molten substance (core fuel) is further heated by the oozy reaction heat, and drops little by little to the containment vessel floor to form a mound. Inside the mound, the mixed molten substance starts to melt by the decay heat. Such a condition continued inside the containment vessel for 10 h from around 4 a.m. on March 12, when the seawater injection started, until 2:30 p.m., when the vent opened.

As the vent opened and the containment vessel pressure stated to drop, the discharge quantity of the fire engine pump increased sharply. As the seawater that dropped to the bottom of the containment vessel accumulates so much as to cover the mixed molten substance, the reaction became active and generated a large amount of hydrogen gas all of a sudden, and the containment vessel pressure increased sharply. This gas pressure lifts up the top lid whose fastening force is weakened. The hydrogen gas that was stuck in the containment vessel escapes through the lifted top lid of the containment vessel into the reactor vault, lifts up the shield plug to flow onto the fuel exchange floor, and mixes with air to form an explosive gas. When the hydrogen flow weakens and the lifted shield plug drops, the impact of the drop ignites the explosive gas to cause an explosion.

That was the study of the core melt and explosion of Unit 1.

Let us list up the new findings and important items.

I. The state of complete loss of water from the reactor pressure vessel continued for at least 4 h. During that period, the core probably collapsed but never reached the state of complete melting. The heat dissipation by radiation continued in the empty reactor pressure vessel, a completely new accident condition that had never been experienced in the world.

II. The core temperature under the radiation heat dissipation process must have been around 2,000 °C, or close to the melting point of the uranium, zirconium and oxygen mixed molten substance, and a thermal equilibrium must have been maintained by the radiation heat. The empty reactor pressure vessel temperature is roughly estimated to have risen to 550–600 °C.

III. A portion of the collapsed core broke the bottom of the reactor pressure vessel and dropped to the floor of the containment vessel to accumulate during the water injection period that lasted about 10 h. Then, it reacted with the seawater whose injection rate had increased sharply as a result of the vent opening, and generated a huge amount of hydrogen gas. The damage by this hydrogen gas explosion was limited to the 5th floor of the reactor building. This confirmed the entry route of hydrogen gas onto the 5th floor and also proved that the ignition source of the explosion was the impact caused by the drop of the shield plug which was lifted up by the hydrogen gas.

Column   Were the Holes Developed in the Containment Vessel as Depicted in "China Syndrome"?

I have heard about a calculation which showed that the molten core of Unit 1 caused the bottom of the containment vessel to melt, leaving only several tens of centimeters before it stopped melting. People often ask me as well if any contaminated water is leaking through holes created by the heat of the molten core on the bottom of the containment vessel, as depicted in the movie "China Syndrome."

I learned about the calculation only through media reports and I have no specific information about it. However, based on my experience, I believe that the particular calculation result is overestimating the depth of the melting erosion. The depth of the melting erosion can be estimated with a certain amount of accuracy so long as we know the temperature, mass, contact area, and decay heat of the molten core, but it can be lead to a mistake if the floor is simply consisting only of concrete: a concrete structure normally has reinforcing steel bars embedded in it.

The heat conductivity of reinforcing steel bars is approximately 50 times larger than that of concrete. When the molten core reaches the depth where the reinforcing bars are arranged, the heat is transmitted to areas far beyond the contact area through the reinforcing bars. I wanted to do a calculation but I couldn't get the detailed specifications of the layout of the reinforcing bars for the containment vessel floor.

If it is an ordinary floor, reinforcing steel bars of approximately 1.2 cm in diameter are arranged in a grid-like pattern of a 20 cm pitch, and then covered by 5 cm of concrete.

Since there is no sense of doing a precision calculation, let me make a ballpark calculation. A 1.2 cm thick reinforcing steel bar with a heat conductivity 50 times larger than that of concrete is equivalent to 60 cm-wide concrete. When the molten core reaches the reinforcing steel bar level, the heat that has been heating a floor of only a 20 cm width will be equivalent to be heating a concrete slab of 20 cm + 60 cm = 80 cm width. This means that the thermal effect will be reduced to one-fourth.

The reinforcing steel bars run not only east and west, but also north and south, so that the heat transfer and thinning expands further. If we consider the heat dissipation from concrete itself, the effect expands further. As a result, the erosion of the floor concrete by the molten core stops at about the depth of the reinforcing steel bars. Actually, to reveal my secret concerning this discussion, I once asked the students of my laboratory to do a similar calculation, that remind me of the result at that time.

It seems that the molten core spread over the floor of the containment vessel only in the case of Unit 1. Moreover, the amount that was spread was only a portion of the core, not the whole core, even in the case of Unit 1. Combining these estimates, I believe that the melting of the containment vessel was quite limited and must have stopped at around the depth of the reinforcing steel bars.

2.8   Conclusion from Studies of Units 1–3

My explanation of the studies of the core melts and explosions of Units 1 through 3 has become a very lengthy one as you can see. For example, many of you may have forgotten what I said about the core melt conditions of Unit 2 already. That exemplifies how complicated the Fukushima accident is. Therefore, let me summarize here the key elements that are common to all melts and explosions.

The first and most important thing is that core melt occur due to heat generated by a reaction between zircaloy and water.

I have said so again and again and have proven this based in connection with the melts of Units 1–3. A reactor core does not melt because of decay heat. Also, it does not melt completely and flow down like a fluid would, as depicted in NHK TV's graphic image.

Let us recall that the decay heat at the time of core melting, which we discussed here, was only about 1 % of the rated output of the typical reactor at most. If we think about it from the other side, a reactor that is operating normally typically converts output of a 100 times that into electricity while cooling the core. The core is not so weak that it melts down by a decay heat of only 1 % or so. To prove it, we saw out that Unit 1, which did not receive even a drop of water, did not explode until over a day afterwards.

"Low-temperature burn" caused by a little bit of decay heat can be cured with proper care. The only thing that can cause an incurable serious burn like a core melt is an acute zirconium-water reaction that melted the TMI core in just 2 min. In order to cause such a reaction, two conditions must be satisfied: zircaloy must be at a high temperature and a there has to be plenty of water. Since a large amount of water was supplied in the case of the TMI accident, the core collapsed and melted all at once.

In case of the Fukushima accident, there wasn't enough water. The reactors had to wait for a long time before enough water accumulated to cause reactions so severe as to trigger core melts and explosions. It was only after plenty of water had accumulated that a severe zirconium-water reaction occurred, leading to a core melt and a hydrogen explosion.

The process by which water was supplied and accumulated varied among Units 1, 2 and 3. That is one reason that makes it difficult to describe the Fukushima accident. That said, in the big picture, I am sure you understand that the reason the core melts occurred was the same as in the TMI accident: a severe zirconium-water reaction. This is a fact that was reconfirmed for the first time in the Fukushima accident. It is tragic that the accident occurred, but the accident will make a great contribution to the safety measures of atomic power generation in the future.

Those of you who are responsible for the future of atomic power generation must not forget two things: that a severe zirconium-water reaction causes a core to melt, and that it generates a large amount of hydrogen. These two things cause nuclear calamities.

A thing that plays an important role in the process of this reaction is the outer skin (shell) of the melted core – most likely formed by the formation of zirconium oxide – that acts like an egg-shell or the bottom of the pan. This fact cannot be discussed without referring to the PCM test and the TMI accident. Without it, the clarification of the Fukushima accident was impossible. I think God has given us humans the diagram of the molten core of the TMI accident (Fig. 1.1) as a hint for understanding the Fukushima accident.

Although I believe that I have not made any big mistakes in my explanation of the core melt, I admit that there must be many minor places in which it is lacking. While I am ready to accept any criticism, I sincerely hope that those who wish to be scientific leaders in this field spend their energy in clarifying the real identity and performance of the egg-shell, because I believe the egg-shell holds the key to the safety measures of atomic energy.

Secondly, I wish to point out that sharp hydrogen generation and the pressure increase caused thereby are the key factors in hydrogen explosions.

Since the zirconium-water reaction occurs rapidly, it generates a large amount of hydrogen gas all of a sudden, which in turn causes a sharp pressure increase. In the TMI accident, the reactor pressure increased from 9 to 15 MPa within 2 min. In case of the Unit 2 and Unit 3 accidents, the containment vessel pressure increased approximately 1 MPa. Although there is no data for Unit 1, I am sure that a similar pressure increase occurred.

This pressure increase pushed up the top lid of the containment vessel whose fastening bolts had been slackened by thermal expansions, while the hydrogen gas that leaked out through the gap produced there filled up the reactor vault, which in turn pushed up the heavy shield plug, and leaked out onto the fuel exchange floor on the 5th floor (ref. to Fig. 2.21).

In case of Units 1 and 3, the hydrogen that flowed into the fuel exchange room became an explosive gas by mixing with air, ignited due to the impact of the shield plug dropping, and caused the explosion in the reactor building.

In case of Unit 2, since the blowout panel provided on the wall of the 5th floor had been dislodged, the leaked hydrogen gas flowed out to the outside of the building through the openings of the panel. The gas was very hot because it was from the molten core whose temperature was over 2,000 °C. I suppose the flow of the hot hydrogen gas flowed like a river stream by itself, never mixing with the room air, out to the atmosphere. Thus Unit 2 caused no explosion.

That is the summary of the hydrogen explosions of Units 1–3.

The third lesson we learned is that there is a chance that a large scale core melt can be avoided if we can inject water properly.

I mentioned in above that a "low-temperature burn" caused by a small amount of decay heat can be cured with proper care.

The fuel rods that are heated into high temperatures by the decay heat are cooled by the steam flow caused by decompression boiling if the reactor pressure is reduced forcibly. Please refresh your memory. In case of Units 2 and 3, the entire core was cooled to about 150–160 °C. If water was injected instantaneously at this very moment, the low temperature zircaloy does not react with water, so that even though the fuel rods could have broken up, the rods would not have melted. Thus no hydrogen explosion could have occurred.

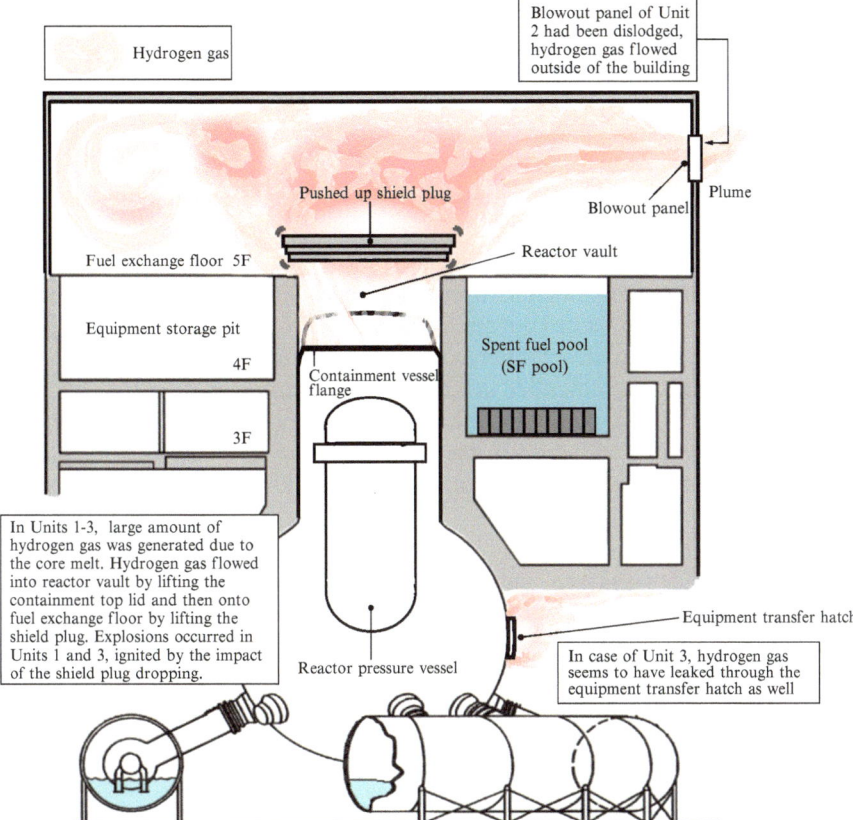

Fig. 2.21 Route of hydrogen gas leakage in case of Units 1–3 (Source: from TEPCO "Fukushima Nuclear Accident Investigation Report")

The countermeasures taken at Fukushima were correct up to the decompression boiling, but the water injection procedure was at least approximately 2 h late in all cases. That was their critical mistake. During these 2 h, the core temperature started to rise again due to the decay heat and the fuel rods became red-hot [IV]. And that's when the water was injected, so that the zirconium-water reaction occurred, leading to the core melt. It was an unfortunate missed opportunity. A delay in countermeasure turned a low-temperature burn into a high-temperature burn.

From another point of view, this means that even when the core water is depleted and fuel is at risk of melting, if decompression and water injection are done promptly and continuously, no core melt occurs even though a core collapse (breakdown of a fuel rod) may occur,. Such a measure will end in a small amount of radioactivity but no hydrogen explosion. If people who work in nuclear power generation remember this and calmly take proper measures with composure, calamities can be avoided.

Fourthly, it is important to estimate the current core status.

People seem to estimate the current status of a core differently.

I estimate that the molten fuels of Units 2 and 3 are all remaining in the reactor pressure vessels and no portion of them has leaked out to the containment vessels. As to Unit 1, a portion of the molten core has leaked out but most of it is still remaining in the reactor pressure vessel.

The reason I think that the molten cores of Units 2 and 3 are remaining in the reactor pressure vessels is that the reactor pressure vessels had some water left when the core melts occurred. Since this is the same condition as that of the TMI accident, I assume that the molten cores of Units 2 and 3 are remaining in the reactor pressure vessels same as in the case of the TMI accident. One of the reasons is that the reactor data after the cores had melted did not change much. I checked the data closely but I could not find any significant difference from those of the TMI.

The Unit 1 core is as described in Sect. 2.7. As described in Appendix 2.11, the γ-ray data at the bottom of the containment vessel measured by TEPCO in the year end of 2,012 showed that the values were larger in the areas closer to the core. This is the reason that I believe that most of the molten core still remains in the reactor pressure vessel.

Then what is the state of the core inside the cores inside the reactor pressure vessels? The TMI data gives us a clue to that question as well. I believe that, as shown in Fig. 2.13, a spherical molten part of about 40–50 cm in diameter is located in the center of an egg-shaped alloy casting of about 3–4 m in diameter and 2 m in height, and radioactive gas continues to emanate from it to this day. The gaseous substances leak out through the cracks formed on said casting, are cooled by the circulating cooling water injected from outside and thus solidify, and are deposited either on the bottom of the reactor pressure vessel or on the floor of the containment vessel.

I think it is safe to assume that said radioactive substances are very fine radioactive particulates. I suppose that some of them have formed water-soluble compounds after causing chemical reactions with water and other impurities. These are considered to have become the source of the contaminated water, as minute amounts of them move along with the movement of water from the openings of the containment vessel to the water sump provided on the underground of the turbine building.

However, we should not attempt to clean up this contamination all at once. Since an abrupt movement of water will take along with it the radioactive substances that have been settled, it could cause a very dangerous situation. It is better to leave it as is without giving it any agitation. It is not the same as environmental contamination issues relating to rain water and underground water.

Since the decay heat is down to less than 200 kW, the radioactive material leaking out of the cracks of the containment vessel is not that much. The temperature of the water remaining in the containment vessel was dropped to 30–40 °C according to a report made at the year end of 2012. Within a few years, there will be no need of cooling by water and the molten part in the middle of the egg will soon be solidified. Now we can safely say that no new incident due to heat can be expected from the radiation emission of the Fukushima accident. The state of the matter as of today has reached such a stage.

Column Possibility of the Molten Core Causing Recriticality

In case of the TMI accident, all the control rods were fully inserted into the core and they melted because the melting point of the control rods were relatively low, and they accumulated in the lower half of the thin outer skin (shell) that surrounded the molten core. Yet still no recriticality occurred in the TMI accident. This fact is evidence that the core never reached recriticality even though the control rods were unevenly concentrated in the molten core.

The core of the light water reactor consists of a large number of fuel rods, each of which is a circular zircaloy cladding tube having a length of approximately 4 m and a diameter of approximately 1 cm, containing numerous fuel pellets, and arranged in an array with a constant interval to form a fuel assembly.

Light water (water) flows around the fuel rods to remove the heat generated by the nuclear fission chain reaction as well as to moderate the neutrons that are generated by the nuclear fission chain reaction and move at high speeds. By the way, slow-moving neutrons tend to cause more nuclear fission reactions. Fast-moving neutrons generated in the fuel are moderated by light water outside the fuel and cause nuclear fissions when they return inside the fuel.

The spacing between the fuel rods is designed to cause the following cycle to occur as efficiently as possible: nuclear fission generate fast-moving neutrons → fast-moving neutrons move to light water → neutrons moderated in light water → moderated neutrons move to fuel → moderated neutrons cause nuclear fissions inside fuel → nuclear fission generate fast-moving neutrons. In other words, keeping an optimum distance between the fuel that generates neutrons and light water that moderates those neutrons makes it easier to cause a nuclear fission chain reaction. This is called the nonhomogenous effect.

When the fuel melts, the light water that moderates neutrons will be removed from the system because of the high temperature, and impurities that prevent nuclear fission reactions will be brought in. Furthermore, the core will become homogenous and lose the nonhomogenous effect that had been prepared with lots of effort to cause nuclear fissions efficiently. As a consequence, the probability of the molten core to reach recriticality becomes very small.

Although the cores and fuel rods of the BWRs in Fukushima are slightly different from those of the PWRs at TMI, their probability of reaching recriticality is as small as that of TMI. Of course, it goes without saying that TEPCO who is in charge of the control is responsible for making sure to provide criticality control for the molten cores by setting up an extreme condition (for example, assuming coagulation by some kind of chemical reaction, etc.) that does not allow any recriticality to occur under any circumstance, but recriticality is not realistic and is something we do not need to worry about.

Appendices

Appendix 2.1

The reactor water level is calculated by the remaining amount of water by subtracting the lost as steam from the reactor. This calculation is a simple one so that we should be able to avoid any error using a computer.

The steam that is generated when the fuel rods are immersed totally in water is all saturated steam. As the water level drops and the fuel rods start to stick out of the water surface, the saturated steam that rises above the water level is further heated to become superheated steam and the temperature rises. It is the same phenomenon as the air being heated when one makes a campfire. As the water level of the core drops, the length of the portion of the fuel rods in the water becomes shorter, so that the water evaporation amount reduces, and the degree of overheating of the steam increases proportionately. In other word, the heat used to be used for evaporation is now used for increasing the steam temperature. Therefore, the temperature of the upper portion of the core rises as well.

Let us now think about the condition where the water level lowers almost to the bottom ends of the fuel rods. Since the lengths of the portion of the rods that are immersed in water are so short, almost no steam is generated. On the contrary, the amount of superheated steam increases and the temperature of superheated steam becomes high and the fuel temperature in the upper portion of the core also rises. The computer does not make any mistake in the computation up to this point.

However, there are few points that are not well thought through in this computation. When the water of the core is lost and replaced by steam, the space occupied by steam increases and the γ-rays generated in the fuel rods disperse to the outside of the core, hence reducing the heating of the core. In addition, the amount of heat that is lost as radiant heat increases dramatically. It seems to me that these two considerations are not sufficiently modeled in the accident analysis computational code. Since the demonstration experiments have not covered that far, there is no proof for them.

Because of their physical nature, γ-rays pass more easily pass through the core which has lost water, enter the reactor pressure vessel and other structural members and generate heat.

Since roughly a half of the decay heat is the heat generated by γ-rays, the calculation done by assuming the decay heat is used for steam generation is overestimating the steam volume to be 20–30 %. I suppose you would agree with me that the more the water level drops, the more a correction is required for such an estimation, if you imagine a condition where the water level drops below the bottom of the core.

Next, the radiant heat. When the fuel rod temperature rises to well over 1,000 °C, we have to think about their radiant heat's effect on structural members of relatively low temperatures such as the reactor pressure vessel. Since the radiant heat is emitted in proportion to the fourth power of the temperature of an object that emits heat, the emission increases dramatically with the temperature. I assume that this radiation heat calculation is not modeled with sufficient accuracy either.

Considering all of these, I assume that, by the time the water level comes down to about the middle of the core, the amount of decay heat transmitted to water decreases dramatically, resulting in the drop of the amount of evaporation, so that the speed of water level drop in the reactor should slow down. However, there is no mention of such computation consideration in the report, and the water drop curve shown in the analysis is straight. Contrary to the evaluation of an imaginary safety analysis, such a correction is mandatory in an accident analysis that pursues the truth.

Appendix 2.2

Assuming the blowout pressure of the safety relief valve is approximately 7 MPa, the specific volume of 7 MPa saturated steam is 0.0274 m^3/kg, while the specific volume of saturated water is 0.00135 m^3, so that the volume ratio of steam vs. water becomes about 20.

As the lowering speed of reactor water (saturated water) level is 4.5 cm/min, the flow speed of the saturated steam is 90 cm/min, i.e., 1.5 cm/s.

Appendix 2.3

Why Did the RCIC Pump Stop?

The question will be answered when the RCIC pump is taken out and disassembled. However, I can guess why.

The reactor water level was maintained at the height of plus 6 m above the core for more than 2 days. This water level is about the same height as that of the steam-water separator. The RCIC pump that had lost the control capability due to the loss of DC power was still operating with the steam energy produced by the decay heat. However, as the water level of the reactor rises unnecessarily, the separation between steam and water deteriorates, and humidity starts to mix in the driving steam. When water mixes in steam, the water drops would pound on the turbine blades just like the sand blasting and damage the blades. If this condition continues, the impeller loses its balance and finally fails. The reason that the reactor pressure started to increase about 2 h before the pump finally stopped seems to be that the pump rotation became abnormal and did neither consume enough steam nor supply enough water.

In the modern technological world we live in, once the mechanical systems lose their control capabilities, they are doomed. As they are then forced to operate under conditions different from their designs, they will suffer malfunctions and damage. Incidentally, while the safety devices such as RCIC were designed based on the premise of 8 h of power loss, the RCIC pump of Unit 2 worked for 3 days. The safety devices worked more than they were designed for. I think this is a proof of excellence of Japanese technology in general, not just nuclear engineering.

Appendix 2.4

Evaporation Amount of Water by Decompression
Boiling (Autonomous Boiling)

At 6:02 p.m. on March 14, the operating staff lowered the core pressure forcibly. The reactor pressure had dropped sharply from about 8 MPa to about 0.4–0.5 MPa. Due to this pressure drop, the water in the lower part of the reactor began violent autonomous boiling. Autonomous boiling is a boiling phenomenon that occurs when the saturation temperature drops with a pressure drop. Let me show you by a ballpark calculation how much water will be lost because of autonomous boiling.

The reactor pressure was about 8 MPa when the safety relief valve was opened. The heat quantity of 8 MPa of saturated water is approximately 1,320 J/kg. If it drops to 0.4 MPa, the heat quantity of the saturated water drops to about 630 kJ. The difference of 720 kJ of heat is to be used for autonomous evaporation. When 0.4 MPa water boils, the heat quantity of the steam is approximately 2,740 kJ. If X% of water is autonomously boiled, the equation is:

$$1,320 = 600(1 - X) + 2,740X$$

so that X=720/2,140=0.3, i.e., it is 30 %. In other words, 30 % of the water remaining in the reactor will be lost by decompression boiling.

As I described in Chap. 1, a BWR pressure vessel is thick and long. There is about 100 tons of water beneath the core. Therefore, it means that 30 % of this, i.e., approximately 30 tons of water, was evaporated.

If I may add a little more, since the pressure is as low as 0.4 MPa, the steam volume increases to about 20 times. The speed with which the steam flows through the core is approximately 0.5 m/s, even if the pressure reduction time is 30 min. This steam flow speed is a sufficient speed to cool the fuel rods.

The core fuel that had been in the sauna bath condition was cooled sharply, although for a short period of time, by autonomous boiling. It requires the help of a computer to calculate accurately how much it is cooled, but I believe that the answer is close to the saturation temperature by the ballpark calculation.

Appendix 2.5

If we assume the hydrogen temperature is 2,000 °C at the core melt time, the weight of the hydrogen gas of 16,000 m³ is:

$$16,000/(22.3/2) \times 293/2,273 = 185 \text{ kg}$$

i.e., approximately 200 kg.

Appendix 2.6

Reactor Water Level Gauge Error After Decompression Boiling

Please look at Fig. 2.22. In "A. Normal operation" shown on the top left corner, we see that the reactor water level can be measured through the pressure difference if the water level inside the reference condensing water chamber is normal.

In "B. After decompression boiling," the reactor's water level is the same as A, but we can see that the pressure difference drops if the water level in the reference level gauge drops such as due to decompression boiling.

In "C. Decompression boiling + water level is below lower piping," the water level in the reference condensing water chamber is the same as in B, the measured water level is not affected by the actual water level change and always indicates a value higher than the actual water level if the reactor's water level drops below the lower side water level piping.

In addition to those above, if the decompression is significant, the weight density change of the reactor water cannot be neglected and should be incorporated.

Appendix 2.7

Zirconium-Water Reaction Where There Is Insufficient Water

Let me explain how I imagine zirconium-water reaction takes place when there is an insufficient amount of water – the so-called "oozy reaction."

The sodium-air reaction that gave me a hint was a reaction between liquid and gas. When the surface film was broken by the generated gas, showing flames, the outside air infiltrated into the inside of the film via the crack. The moisture content contained in the air reacted with sodium and produced hydrogen, which in turn became the gas to cause the next crack. This slow reaction continued again and again.

The oozy reaction is essentially a reaction between high temperature zirconium and water. When a red-hot fuel rod is broken into pieces when it meets with cold water, hot zirconium agglomerate (or liquid) comes into contact with water to generate an enormous amount of oxidation reaction heat locally, which in turn melts uranium dioxide in the vicinity or produces an alloy with zirconium, and causes the fuel pellets that were originally discrete pieces to weld together. During this process, the surface in contact with water must have formed a zirconium oxide coating, even though it might not be perfect.

I believe that this imperfect coating increased its size each time the reaction occurred, and eventually produced a shell like a cooking pot that protects its contents from coming into contact with water. This is the initial stage of the oozy reaction and I suspect that it was most likely formed on the core plate. Therefore I believe that the molten substance contains lower melting point stainless steel as well.

As a relatively large amount of heat was generated by this first reaction with water, the water level dropped by evaporation and I suppose that the pot contained

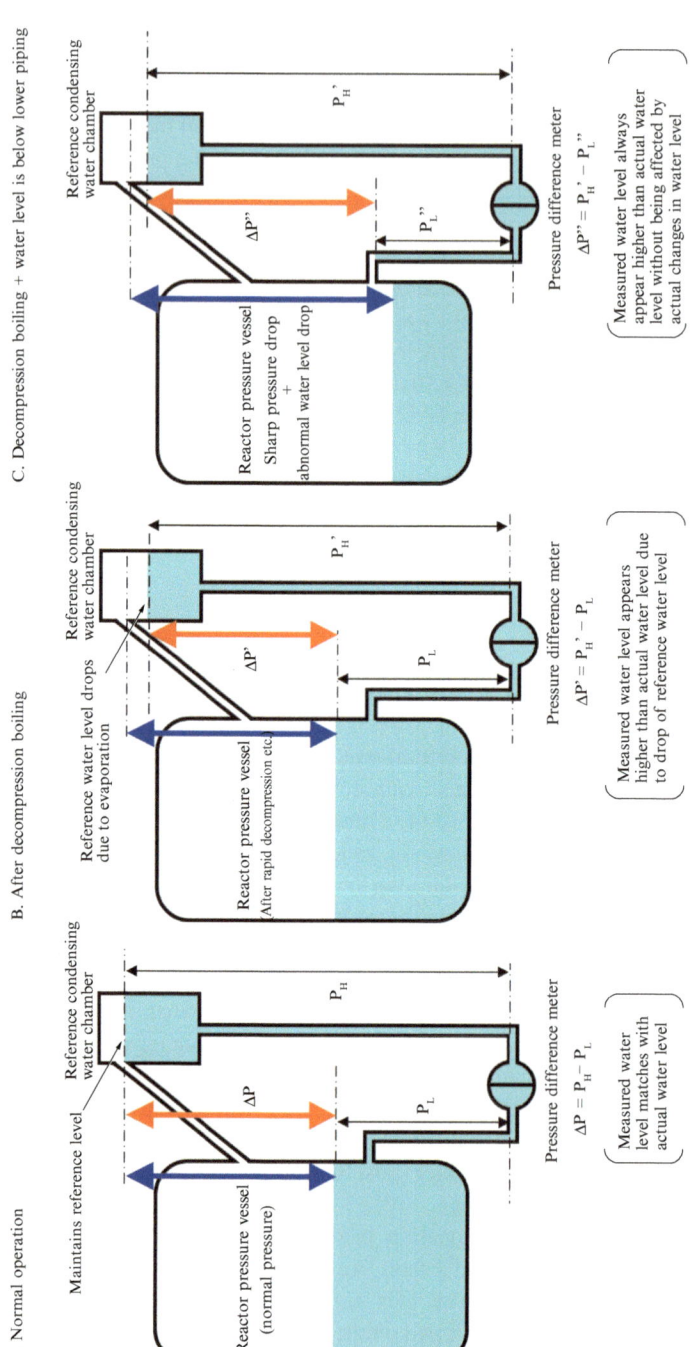

Fig. 2.22 Reactor water level gauge error after decompression boiling (explanatory diagram)

molten fuel rods and other things to a certain degree. The surface of the molten substance was covered by a thin oxide film and a small amount of reactions must have continued as the surrounding steam flowed in each time hydrogen gas was discharged. This is the oozy reaction. I believe that, during the oozy reaction period, the majority of the fuel rods must have stood upright on the pot made of the molten substance although they may be slightly bent or broken, and that the core as a whole had not collapsed.

The pot acts as a shield against heat. The fuel rods in the pot are heated by the decay heat and the heat of the oozy reaction and gradually melted from the lower part. As a result, the portion of the core exposed above the water was eventually swallowed by the rest as time passed.

On the other hand, the water level of the coolant water recovered over time, eventually exceeding the edge of the pot. The second reaction started in such a condition.

In addition to the fact that the temperature of the mixed molten substance itself had become higher, the water had increased so much as to go over the edge of the pot. As a result, the second reaction was huge, and melted the core.

That is what I believe happened in the oozy reaction.

Appendix 2.8

Temperature Calculation Related to Radiation Heat

An accurate calculation of radiation heat can be very complicated. However, its feature is that the heat exchange is conducted in proportion to the difference between the 4th power of the absolute temperatures of the heat-emitting and -receiving objects.

Let us assume the surface area of a mixed molten substance with a temperature of 2,000 °C (2,270°K) is 1 (half of the normal core surface, assuming that it is collapsed to a certain degree), the total surface area of the core shroud covering the core, upper part core plenum, and the top guide and core plate is 4, and the total surface of the reactor pressure vessel (approximately 20) and other internal components is 50. We will now make a rough calculation, disregarding the surface absorption rate and radiation shape factor, and assuming that the heat among the core, shroud and pressure vessel have reached an equilibrium state.

Paying attention to the fact that the radiation heat is proportional to the 4th power of the absolute temperature, and assuming the temperature of the shroud etc. is $A°K$, the equilibrium equation is as follows: since "$2,273^4 \times 1 = A^4 \times 4$," the equilibrium temperature is "$A = 1,600$ K (approx. 1,370 °C)."

The temperature B of the reactor pressure vessel is: since "$1,600^4 \times 4 = B^4 \times 50$," "$B = 850$ K (approx. 570 °C)."

Appendix 2.9

As the initial assumption for the calculation, we assume the temperatures of both the reactor pressure vessel and the core internal structural members are 250 °C.

Assuming that they are all made of steel, we assume the specific heat of 0.54 kJ/kg °C, the melting point of 1,500 °C, and the heat of solution of 272 kJ/kg.

The heat quantity required for the reactor pressure vessel (340 tons) and the lower half of the internal components (50 tons) to be heated to 550 °C is:

$$(340+50)\times10^3\times(550-250)\times0.54 = 6.3\times10^7\,\text{kJ}\,(17\,\text{MWh})$$

The heat quantity to melt the upper half of the core internal structural members is:

$$50\times10^3\times(1500-250)\times(0.54+272) = 4.7\times10^7\,\text{kJ}\,(13\,\text{MWh})$$

It will require a heat quantity of approximately 30 MWh.

Appendix 2.10

Asking the fire station about the typical shutoff pressure (pump pressure when the discharge quantity is zero) of a typical fire engine, I got the following answer:

"Although it varies with vehicle types, the typical pump vehicle's pump discharge pressure is approximately 0.85 MPa at most, and the nozzle pressure is 0.4 MPa. The discharge rate is 24–40 cubic meters per hour."

We do not know how the fire engine was connected at the time of the accident, but since the containment vessel pressure during the time period before the vent was opened was 0.8 MPa, it was close to the pump shutoff pressure. There is no question that the pump discharge rate was very small. When they opened the vent, the containment vessel pressure dropped to around 0.5 MPa so that the seawater injection rate increased substantially.

Since the flow resistance up to the nozzle varies with the length of hose connected to the pump, the discharge rate varies with it. Since the length of the hose cannot be assumed to have been as short as that when fighting fires, we guesstimate that it was 30 m³/h. This makes the time until explosion to be around 1 h.

That means that the remaining 50 m³ of seawater was pushed over a period of as long as 10 h, fighting the pressure of 0.8 MPa, which was close to the pump shutoff pressure (pump pressure when the discharge quantity is 0). The discharge quantity was 5 m³/h in this case.

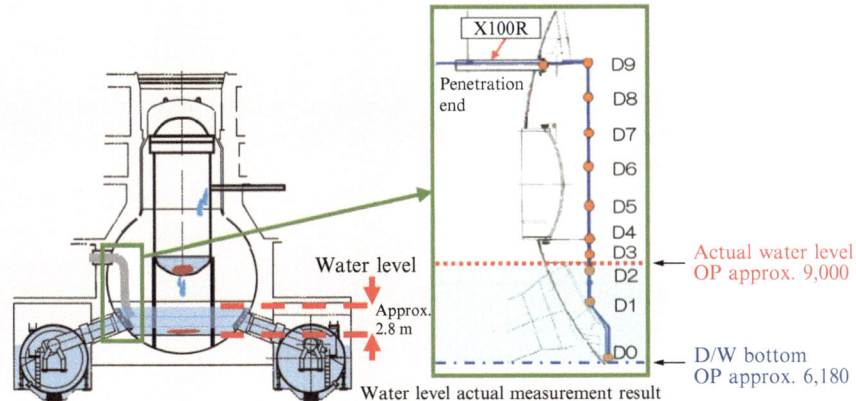

measured radiation level

Measuring point	Height from D/W bottom	Radiation level data (Sv/h)
Penetration end	8,595	Approx. 11.1
D9	8,595	9.8
D8	Approx. 7,800	9.0
D7	Approx. 6,800	9.2
D6	Approx. 5,800	8.7
D5	Approx. 4,800	8.3
D4	Approx. 3,800	8.2
D3	Approx. 3,300	4.7
D2-water surface	Approx. 2,800	0.5
D1	—	—
D0	0	—

Fig. 2.23 Unit 1 DW radiation and water level measurement data (Source: from reference materials of "TEPCO Intermediate-Long Term Countermeasure Meeting/Administration Meeting (Dec. 3, 2012)")

Appendix 2.11

Many people seem to believe that the molten core of Unit 1 flowed out of the reactor pressure vessel almost completely. However, I think the majority of it still remains in the reactor pressure vessel.

The reason can be found in Fig. 2.23, the result of the measurement made on December 3, 2012 by TEPCO as a preparation for the decommissioning measures. As the figure shows, the radiation level of the lower part of the containment vessel reduces to one-third as the measuring point lowers from D-9 (a point close to the bottom of the reactor pressure vessel) to D-3 (approximately 5 m below D-9), in other words, as the measuring point moves away from the bottom of the reactor pressure vessel.

The radiation is stronger at the bottom of the reactor pressure vessel than at the lower part of the containment vessel. This proves that the molten core still remains in the reactor pressure vessel.

However, we are not sure how much remains. Judging from the course of events, as well as from the temporal speed of the egg-shell formed by the mixed molten substance, I assume that about a half of it still remains inside. The above was written based on that assumption.

Reference

1. Ishikawa M (1996) Reactor power excursion [Japanese version only]

Chapter 3
Fukushima Daiichi Unit 4 Accident

Unit 4 is a sister unit of Unit 3. Its control room is located on the same floor as the control room of Unit 3 with no partition between them. Its operation started in October 1978. On the day when the earthquake occurred, all of its fuel assemblies had been taken out of the reactor in order to replace the core shroud for the purpose of preventive maintenance against stress corrosion cracking and had been placed in the spent fuel storage pool (SF pool), which is provided adjacent to the containment vessel. This itself developed into another big issue unrelated to the core melt and explosion, which are the main subject matters of this book.

Just incidentally, I visited Fukushima Daiichi and watched the shroud replacement work on Unit 4 the day before the day of the earthquake, March 10, 2011. I saw the entire fuel exchange floor was occupied by all kinds of reactor-related components including the top lid of the reactor, the top lid of the containment vessel, and the shield plug. The top of the reactor was filled with water for the core shroud replacement work and its water level was at same level as that of the SF pool across a partition. The site was full with many workers and it seemed that the work was going on without a hitch. I had no idea at all at that time this observation would later be useful, as I will describe later on.

Now, there are two things to talk about when it comes to Unit 4. One is the explosion in Unit 4 building, which is an extension of the studies discussed in this book so far, while the other one is the sensational uproar created by concern expressed in the United States as a result of said explosion – a concern that the explosion might be a result of a melting of the spent fuel rods caused by the SF pool of Unit 4 being damaged by the earthquake and thus losing water. For about 10 days immediately after the explosion of Unit 4, the latter, rather than the core melts in Units 1–3, was a matter of global concern.

In reality, the explosion of Unit 4 was caused by the back-flow of hydrogen gas generated in Unit 3, as explained in Chap. 2. However, this came to be known at the end of August, about 5 months later after the accident. It was found out that the degree of contamination of the filters attached to the Standby Gas Treatment

© Springer Japan 2015
M. Ishikawa, *A Study of the Fukushima Daiichi Nuclear Accident Process*,
DOI 10.1007/978-4-431-55543-8_3

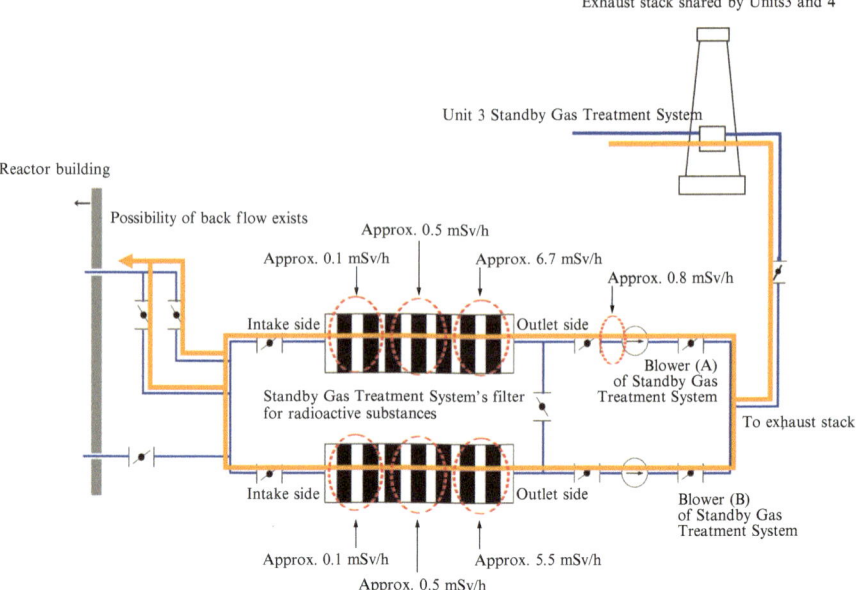

Fig. 3.1 Contamination condition of Unit 4 Standby Gas Treatment System's filter for radioactive substances (Source: from TEPCO "Fukushima Nuclear Accident Investigation Report")

System (SGTS) of Unit 4 was heavier at the outlet side than at the inlet side, which normally would not happen (Fig. 3.1). This proved the back-flow of hydrogen from Unit 3 to Unit 4.

For the several months until then, most of those specialists around the world who are associated with nuclear power generation believed the speculation that the explosion of Unit 4 was caused because the SF pool was damaged by the earthquake, causing water to leak, which in return caused the spent fuel rods to melt in the emptied SF pool because of the decay heat, and hence caused the hydrogen generated by the zirconium-water reaction to explode in the end. This speculation was widely spread into the world with strong credibility.

This was a typical example of harmful rumors. Even some of my close American friends repeatedly asked me by email whether we had yet to find the leaks from the Unit 4 SF pool. In fact this stream of email continued until the early autumn. That is how mysterious the Unit 4 explosion was.

The explosion of Unit 4, which was idling for a periodical inspection, was mysterious for everyone. No one could provide a convincing explanation at the time. The reason that the United States issued a recommendation for U.S. citizens to stay away from a zone with a radius of 80 km around Fukushima Daiichi was that they were concerned that the spent fuel would melt and a large amount of radioactivity could be released like at Chernobyl if the SF pool was damaged.

The explosion of Unit 4 attracted the world's attention because its cause was not clear, and made not only US nuclear specialists but also those around the world worry about it. They claimed that "Japan is hiding," "not telling the truth," "denied the facts of core melts until the end of May," and "reported that the accident level is 4 in the beginning, raised it to 5 and finally to 7." It has only itself to blame, but the government of Japan was that untrustworthy in the eyes of foreign countries.

The eyes of TEPCO's staff members which focused their attentions on the contamination distribution solved the harmful rumors and the distrust of Japan by foreigners in one stroke.

3.1 Explosion of Reactor Building

It was around 6:14 on March 15 when the explosion occurred in Unit 4. A TEPCO report says only that there was a big shock sound and vibration, and then damage were noted later in the vicinity of the 5th floor roof of the reactor building. It does not explicitly say that it was an explosion. The expression is slightly ambiguous.

This is because the containment vessel of Unit 2 is estimated to have been damaged at about the same time. As an evidence, the radiation level in the vicinity of the front gate started to rise sharply about the same time as shown in Fig. 2.4.

Two incidents occurred almost simultaneously. Therefore, TEPCO probably decided not to make a decisive comment as both are incidents that must have accompanied some shock sounds. They claim that they heard a shock sound when they were making the morning patrol. Judging from the description that a shock sound was heard, I estimated that they must have heard the shock sound, not from the damage of the containment vessel but from the explosion of Unit 4, whose energy was larger than the other.

The TEPCO's report says that a fire occurred in the north-west corner of the 3rd floor of the reactor building of Unit 4 at 9:38 a.m. on March 15, i.e., after the explosion, and the fire died automatically at around 11 a.m., while flames were noted in the north-west corner of the 4th floor at 5:45 a.m. on March 16, but no flame was noticeable at 6:15 a.m. of the same day. I suppose both of these foxfires were the burning of the hydrogen gas left after the explosion.

The explosion of Unit 4 was caused by the hydrogen gas that was generated due to the core melt in Unit 3, then flowed back along the duct of the Standby Gas Treatment System of Unit 4, and entered into the reactor building as explained in Chap. 2. Therefore, I believe that the amount of hydrogen gas itself was not that much and it was a relatively low concentration hydrogen gas, as it was mixed with air and became thinner in the course of its back flow. I suspect that this hydrogen gas gathered around the ceiling area of the reactor building because of its lightness, and became thick enough to be an explosive gas.

The mechanism of the flow of hydrogen gas from Unit 3 to Unit 4 is that the hydrogen gas from Unit 3 caused the pressure in the narrow space of the lower part of the stack to increase, developing a pressure difference between there and the reactor building of Unit 4, which was in an atmospheric condition, and hence caused a back flow by the pressure difference. There is no doubt that there was a back flow, judging from the contamination condition of the filters. I would not have believed this theory had this evidence not been available. It is that rare a phenomenon.

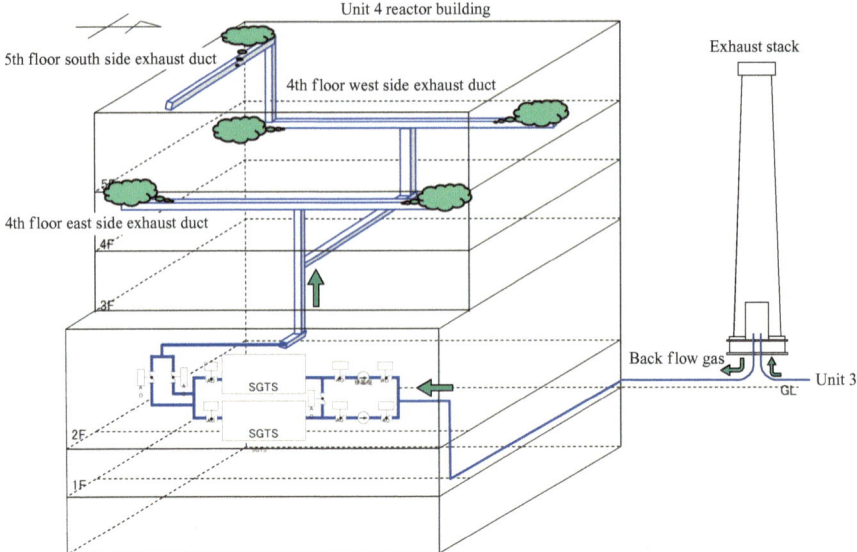

Fig. 3.2 Hydrogen gas passage from Unit 3 to Unit 4

It is against general common sense of engineering for hydrogen gas to flow back through the filters, which have high resistance, due to the small pressure difference between the lower part of the stack and the reactor building of Unit 4. In order to understand the delicate pressure difference in this area, one must understand thoroughly the specific structure of the lower part of the stack and the characteristics of the filters. There is no doubt that hydrogen flowed from Unit 3 to Unit 4 as TEPCO explained (Fig. 3.2).

As it was already explained in Sect. 2.6.5 of Chap. 2, there can be two ways in which this particular hydrogen gas flow might have taken place. One of the possible flows is from the vent of Unit 3. The theory is that the pressure of the space of the lower part of the stack increased as a result of the pressure of the hydrogen gas discharged from the vent, and the gas flowed into Unit 4 via the high-resistance filter.

The other theory is that since there was a certain amount of leakage in the containment vessel of Unit 3 – I suspect that it was through the seal of the equipment transport hatch – hydrogen gas continued to leak into the reactor building of Unit 3, then eventually reached the bottom part of the stack via the duct of Unit 3, and then slowly sneaked into Unit 4. Since it is safe to assume that there was approximately 26 h for the gas to leak in this scenario, so long as there was even a slight pressure difference, a constant flow would have existed.

In any event, the pressure of the lower part of the stack did not have to be that much, as what is required is only just enough pressure to move the light gas up to the upper level of the stack. The pressure of the hydrogen gas that accumulated in the lower part of the stack was probably slightly higher than that in the Unit 4 building.

The hydrogen gas, in accordance with the pressure difference, must have slowly flowed backward through the ventilation duct into the reactor building of Unit 4.

The light hydrogen gas must have entered the reactor building after leaving the duct, dwelt in the ceiling area of each floor, and increased its concentration as time went by. Since it is difficult to assume that the amount of hydrogen gas that entered was so much, I imagine that the mixture with air was not that dense; rather, it must have been a density barely reaching the explosion threshold.

The difficult part is how this hydrogen gas was ignited. This time there is no shock of the dropping of the shield plug as in the cases of Units 1 and 3. No tremor or impact due to earthquake was recorded at the time of the explosion. While I was perplexed with the mystery, I got an interesting hint. It was that the investigation by TEPCO after the accident reported that a wire mesh provided at the side of the SF pool of Unit 4 was found deformed protruding toward the SF pool. Since this wire mesh is provided at the air suction port, it can be deformed concavely, but it is never expected to deform convexly like that.

In addition, I heard from TEPCO staff during my visit to the explosion site of Unit 4 after the accident that the explosion was more likely to have occurred near the ventilation duct of the 4th floor. The TEPCO's final report also mentions that the duct that existed on the 4th floor was completely shattered. I believe that this view of TEPCO's is right on the dot.

The difference of Unit 4 from Units 1–3 is that all the fuel rods were removed from the core and placed in the SF pool. It is said that the decay heat of the SF pool of Unit 4 was 4 times higher than the other SF pools because of it. The temperature of the SF pool that lost the means of cooling due to the power failure naturally rose. According to a calculation, the temperature of the SF pool of Unit 4 reached the saturation temperature of 100 °C by 3 days after the accident and the water level was gradually dropping as the water reached the boiling point.

I suppose that the other side of this fact means that the Unit 4 reactor building must have been completely filled with saturation steam[1] produced from the SF pool water by March 15, the day the explosion occurred. In other words, it was in a condition of a sauna bath of nearly 100 °C temperature. This naturally means all the sub-assemblies and equipment placed in the building was heated as well. Especially the ventilation ducts made of thin steel plates must have been heated close to 100 °C. A duct extends due to thermal expansion as its temperature rises.

Under hot summer weather, railway rails can be heated to high temperatures by direct sunlight, bend due to thermal expansion, and in the worst cases cause derailment accidents sometimes. That is why a gap is provided between the two adjoining rails – that is to avoid deformation due to thermal expansion. That is also why we hear those regular rhythmic sounds when railroad vehicles go over the gaps. Without this noise, the train may derail.

However, we cannot provide a gap in a ventilation duct. If we provide a gap, the suction or discharge air that goes through the duct would leak. Instead of providing

---

[1] Steam generated by evaporation of water. The temperature of steam is uniquely determined by its pressure. For example, the temperature of steam of 1 atmospheric pressure is 100 °C.

a gap, a duct is normally used under a relatively constant temperature condition. Since the temperature abruptly rose to 100 °C in our case, the duct must have deformed due to thermal expansion.

The fuel exchange area on the 5th floor is a long room having a large traveling crane. The ventilation duct is installed in a straight line parallel to the crane to avoid conflict with the crane movement. Therefore, the duct has little constraints and can extend due to thermal expansion rather freely. However, the situation is entirely different on the 4th floor. The route of the duct on the 4th floor is very complex, going up and down, bending like a crankshaft or branching out in some areas in order to avoid conflict with various equipment installed on the floor. In addition, the duct is constrained with ties provided at various locations so that the duct cannot extend freely and may develop distortions due to temperature changes. Since the duct is made of thin steel plates, it does not have much strength. If it is subjected to a large strain, it may buckle or break. This buckling and breaking can be the ignition source of a hydrogen explosion.

The mode of air heating in this case, where the room temperature is affected by the evaporation (boiling) from the surface of the SF pool, is such that a hot air layer is formed at the top of the room and the boundary of the hot air layer comes down gradually, rather than the hot air randomly mixing with the room air. More specifically, a hot air layer in the reactor building is first formed near the ceiling of the 5th floor, then the entire 5th floor is gradually warmed up, and finally the 4th floor is warmed up from the top.

The morning of March 14 was about the time when the air warmed up to around 100 °C came down to the 4th floor. As the layer of heated air descended, thermal expansion and buckling occurred in the 4th floor duct. The hydrogen inside the duct was ignited by this buckling.

The hydrogen explosion probably did not occur in the 4th floor duct; rather hydrogen combustion must have started in the duct. That is my guess based on the fact that the wire mesh at the suction port on the side of the SF pool was protruding. This hydrogen combustion exited from the duct and traveled to the dense hydrogen gas that had accumulated at the ceiling to cause an explosion first on the 4th floor. Its impact then jumped to the hydrogen gas accumulated at the ceiling of the 5th floor to cause the explosion of the reactor building of Unit 4. It must have happened within a very short period of time. The explosion essentially started on the 4th floor and jumped to the 5th floor. It is opposite to the sequence that occurred in Unit 3.

At the explosion site of Unit 4, there are signs that the explosion pushed up the floor of the 5th floor and dented the floor of the 4th floor. It must have been the hypocenter. The 5th floor ceiling was destroyed and the walls had fallen down as if they had been pushed outward while maintaining their shapes. Since hydrogen is light, it probably accumulated in areas near the ceiling, increasing its density and exploded. The reason that the walls were pushed down was probably because the explosion force pushed the top of the walls.

I personally experienced air raids three times during the war. The first experience was that of a bomb and the other two experiences were those of incendiary bombs. Japanese wooden houses were defenseless against incendiary bomb attacks and experienced major damage by fires, while the conventional bomb attacks caused less effects. While the conventional bombs had a deadly effect on the houses that were hit directly, they did not affect the surrounding houses much. Although this is all from my childhood memories so that I am not too sure, I remember hearing that they were 500 kg bombs. The bombs typically made conical holes that were 6–7 m wide and 3 m deep on the ground. The strength and effect of the explosion of Unit 4 were incomparably larger than those of the bombs. Metaphorically speaking, it is like comparing a lizard to a dinosaur.

I once heard from a visiting specialist that "hydrogen explosions are severe and horrible; they are hardly comparable with steam explosions" when I was studying excursion phenomena of nuclear reactors in U.S., and I found out that he was correct. And yet, the explosion of Unit 4 is much smaller than that of Unit 3. The most severe hydrogen explosions ever recorded in the history of mankind were most likely Chernobyl and Unit 3 of Fukushima Daiichi.

On a different note, let me give you an interesting episode. The specialist I mentioned above told us that " hydrogen explosions run horizontally." When I asked him if he experimented it, he replied "I have never done such a dangerous experiment." He said that it is only hearsay. However, I found out that his story was correct. I suppose that a lot of people saw a spark run sideways from the ceiling of the building in the TV images of the Unit 3 explosion. "Hydrogen explosions run horizontally." I don't know who but our forerunners must have done experiments to leave us with even such a finding.

I visited Unit 4 of Fukushima Daiichi after the accident to see its explosion site. I stayed there probably only 30 min or so. Although the site had been tidied up somewhat to allow us to see the site, the route from the 5th floor to the 1st floor that snaked through the explosion debris, circled around the reinforced SF pool, and followed temporary catwalks to reach the 1st floor was really tough, just like an obstacle course. The obstacle race played with a full-face mask was a severe challenge to this old man who was almost 80 years old, but the total exposure dose I received only 0.1 millisievert. Moreover, the most of it was what I received from Unit 3 while I was standing on the wall-less 5th floor. I was told that the radiation level of the 5th floor is roughly 0.3–1 millisievert and the radiation from Unit 4 is very little.

3.2 Global Influence of the Explosion

The explosion of Unit 4 caused an immediate concern for the United States. It was that the explosion of Unit 4 may be the recurrence of the Chernobyl accident, which emitted a large amount of radioactive materials to the environment.

The concern emanated from the question of why an explosion occurred in the reactor building, where the fuel rods were all taken out from the reactor for the periodical inspection. Nobody knew on March 15, the day the explosion occurred, that the hydrogen gas from Unit 3 had seeped into Unit 4. First of all, it is quite rare in nuclear power plants for two reactors to share one stack, so that nobody imagined that hydrogen gas would sneak in from the adjacent reactor.

In Unit 4, where they were conducting a periodical inspection, all the fuel rods in the reactor were taken out and temporarily placed in the SF pool. Since it was heated by the decay heat from fuel rods that had been taken out not long ago, the heat of the SF pool of Unit 4 was almost four times larger than those of other pools.

The U.S. Nuclear Regulatory Commission supposed that the explosion of Unit 4 must have been caused by their understanding that the SF pool lost all its water as a result of the earthquake damage, hence the recently removed fuel rods which were still experiencing high decay heat got very hot, and thus generated hydrogen gas as a result of the reaction between the zirconium alloy of the cladding tubes and the water vapor in the atmosphere.

As the SF pool is outside of the containment vessel, there is nothing to cover the fuel rods if the reactor building explodes. They seemed to have thought that if the fuel rods had melted, it would become the same as Chernobyl where the core was damaged and exposed to the outside, so that the radioactive emission would soon start from the molten fuel rods lying under the open sky.

As a countermeasure, the U.S. government issued an order restricting U.S. citizens from entering the area within 80 km of the power plant. This was a reasonable decision at the time, when the cause of the explosion was unknown.

The U.S. was not the only country which was afraid of damage caused by radiation. It is said that most of the staff members of embassies located in Tokyo left Tokyo hearing the melting and explosion of Unit 1. Foreign mass media staff members located in Tokyo whose business is to collect news were no different.

In the meanwhile, the U.K. was unique. It is said that the Government Chief Scientific Adviser of the U.K. told the entire staff members of its embassy in Tokyo that no person living in Tokyo would be affected by radiation, as it would be impossible for a large contamination to spread outside of an area of 30 km radius from the explosion site judging from a calculation based on the height of the explosion smoke of Unit 3. His words proved to be true.

Japan could not respond properly to the concerns of the United States. If they were calm enough, it was simply a matter of dispatching staff members to check if there was any water in the Unit 4 SF pool. Even if the radiation level was high, it was only a status check to be completed in a short period of time, so that it could have been completed without causing any physiologically harmful exposure. In fact, TEPCO staff members measured that the SF pool water temperature was 84 °C the day before the explosion. If they implemented the check, the concerns of the U.S. could have been solved on the spot. There was no need for an extravagant operation, like the seawater spraying operation by the Self-Defense Force.

It was perhaps futile to have asked for leaders, who had become so hot-headed, to be level-headed. In response to the concerns expressed by the U.S., the government ordered a JGSDF helicopter to spray seawater on the SF pool of Unit 3 that caused the explosion. Also on this day, Prime Minister Kan requested the Metropolitan Government of Tokyo to dispatch fire engines of the Tokyo Fire Department that are designed to fight skyscraper fires. Shintaro Ishihara, the governor of the Metropolitan Government of Tokyo, responded to this request, and an attempt to inject water to the SF pool was started on March 19. An extensive operation such as this was conducted at the accident site, rubble-strewn by the tsunami and explosions and contaminated by radiation.

These failures in judgment added further complications to the already confusing accident site, and the news communicated by the mass media brought dismay after dismay for the citizens, while the lack of sensible explanations on the main issues, i.e., the reactor accidents, simply increased the uncertainties in the minds of people.

In actuality, at the time when these operations were ordered, the SF pool still had sufficient water and so that there still was ample time before water injection became really necessary. In essence, it was not quite necessary to carry out the extensive operations on Unit 4 as such, when they were busily responding to the accidents of Units 1–3. However, I truly take my hat off to the courage of those who risked their lives in this operation.

In the end, people around the world let out a big sigh of relief as they heard the news that sea water was securely injected into the pool with the help of a concrete pump vehicle, nicknamed as "giraffe," which joined the operation on March 22.

At the time of the accident, the reactor of Unit 4 was filled with water for the core shroud replacement work and all the fuel rods were moved to the SF pool. The water filled up to the top of the core was separated by a partition between the SF pool, and the partition was designed to give in to the pressure to allow the water to flow from the reactor side to the SF pool in case the water in the SF pool decreased. In other words, the SF pool had a water source next door, which is capable of replenishing water in case of need. The U.S. government was not aware of this.

We all saw on TV the video image taken when the TEPCO staff members who rode the JSDF helicopter on March 16 after the accident checked the SF pool from above. The SF pool filled with blue water was seen on the video, peeping through the holes made on the walls by the explosion. If they had observed the image with calm minds, they should have been able to see that the SF pool had not lost water.

Let me sidetrack again with my personal episode, but I went to Washington D.C. on March 23, 2011, which was immediately after the accident. Since I belonged to the Japan Nuclear Technology Institute, which was a position of leading the nuclear energy industry of Japan, I felt the responsibility to report about the current situation of the Fukushima accident to the cooperating associations of the U.S. and other leading industrialized nations, and made the trip voluntarily. My effort was very much appreciated by the people whom I met with as I presented the information I got through the newspapers in an organized manner unit by unit, although the information was quite limited. The people overseas were very much confused at the time as the melts and explosions had occurred in as many as four reactor units.

To my surprise, the other group of people who welcomed us was the Japanese journalists in Washington, D.C. They were not receiving any information. On the other hand, their foreign correspondent friends were relying on them as they suspected that the Japanese media people must be well informed. The Japanese media people there were very much upset because they couldn't get any information. It sounded like our information was a gracious rain to them.

But I digress. The concern of the U.S. NRC at the time when I met them was on the soundness of the SF pool of Unit 4, rather than on the accident information of Units 1–3. As I told the fact that I was at the site of Unit 4 on March 10, the day before the earthquake, that I saw water in the SF pool in the video image taken from the helicopter, and that the SF pool of Unit 4 had water for sure, they looked very relieved, and the NRC Chairman (incumbent at the time) Gregory Yaczko who was out at the time came back to the office specifically to express his special appreciation for our visit. Although I am not sure if it is related to their appreciation, it was on March 25th that the "Operation Tomodachi," the support activities by the U.S. government to Japanese people suffering from the Fukushima disaster, picked up the pace.

I visited cooperating civilian organizations in the U.S. and other leading industrialized nations at about the same time, and found that their concerns were also on the soundness of the SF pool of Unit 4 in addition to the core melt and explosions of Units 1–3. In other words, the cause of the explosion of Unit 4 was that mysterious to everybody.

Part II
Improving Nuclear Safety and Reconstructing Fukushima

Chapter 4
Release of Radioactive Materials and the Evacuation of Residents

4.1 Background Radiation Level Increased by the Release of Radioactive Materials

First, Fig. 4.1 shows changes in radiation level in the vicinity of the main gate of the plant. This graph is generally the same as Fig. 2.4 except for some annotations. The main gate, around which a radiation measurement instrument is installed, is located about 1 km west of the four-unit plant.

Look at the entire figure aside from the numbered rapid increases in radiation level. The background radiation level increased twice, as indicated by two plateaus in the chart, which is the most distinct feature of this figure.

At around 4:00 a.m. on March 12, the day after the accident (number ①), the radiation measurement instrument indicated an increase in radiation from about 0.07 to about 4 μSv/h. Subsequently, radiation was released on a massive scale with two peaks (⑧ and ⑨) throughout almost the whole day from around 10:00 p.m. on March 14, whereupon the background radiation level reached up to around 300 μSv/h at around 8:00 p.m. on March 15 and stabilized. In this manner, the release of radioactive materials from the Fukushima Daiichi NPS is characterized by two increases in the background radiation level. How these changes in background radiation level took place and how they affected the evacuation of local residents are the main subjects of this chapter.

You can see that, based on a comparison between the changes in the background radiation level and the accident situation, the initial increases in the background radiation level from March 12 until late at night on 14 resulted from a radioactive release due to the core melts at Units 1 and 3, while the subsequent increases late at night on March 14 resulted from a direct radioactive release from Unit 2.

The term "background radiation level" I have used means the baseline radiation level of a place. For example, following the rapid radiation level increase (②) at 10:17 a.m. on March 12, the radiation level immediately decreased to 4 μSv/h. I have described the stabilized radiation level 4 μSv/h using the term "background

© Springer Japan 2015
M. Ishikawa, *A Study of the Fukushima Daiichi Nuclear Accident Process*,
DOI 10.1007/978-4-431-55543-8_4

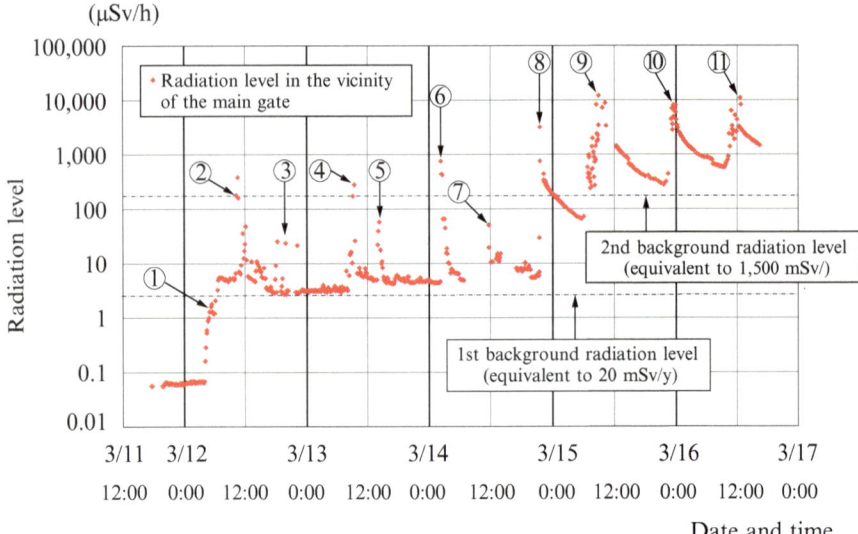

Fig. 4.1 Changes in radiation level in the vicinity of the main gate of the Fukushima Daiichi NPS (measured values) (Source: based on "The Fukushima Daiichi NPS Accident Investigation Report" by the TEPCO)

radiation level." This can be interpreted as the representative radiation level in the vicinity of the measurement point.

Each of the numbered rapid radiation level increases is discussed in Part I, Chap. 2 concerning core melts and explosions. Also in this chapter, I will explain the reasons for the increase in an integrated manner at the end of this section.

4.1.1   First Increases in the Background Radiation Level – Amount of Radiation Released from Units 1 and 3

First, with regard to increases in the background radiation level at around 4 a.m. on March 12, I did not clearly understand the reason at first.

In Part I, Chap. 2, I explained how the increases in the background radiation level were caused by radioactive releases by venting the SCs at Units 1 and 3. Frankly speaking, this explanation is incorrect. This explanation was intended to concentrate on discussing the complicated phenomena of core melts and hydrogen explosions by omitting a bothersome explanation on reference radioactivity data and simply describing that the radiation level increased due to vent. It was intentionally omitted for the purpose of convenience to reduce the complexity of Part I, Chap. 2. I correct the previous explanation here.

Next, I move on to discuss increases in the background radiation level ①. At 4 a.m. on March 12 when the background radiation level soared (①), the Unit 1 SC vent had not yet been opened. Accordingly, the increase in background radiation level to 4 µSv/h is not attributable to the SC vent at all. The SC vent was fully opened at around 2:30 p.m. on March 12, rapidly reducing the containment vessel pressure. This rapid decline in pressure caused seawater from a fire engine to enter the reactor vessel, resulting in a hydrogen explosion, as I detailed in Part I, Chap. 2, Sect. 2.7.

What caused the radioactive release to increase the radiation level at 4 a.m.? This question remains unanswered. The only work performed at the time was to initiate the water injection at Unit 1. Here I outline a report on this period provided by TEPCO.

"A water inlet was found behind the protection door of a turbine building entry-way at around 3:30 a.m. on March 12 and at around 4 a.m., fresh water loaded on a fire engine was injected into the water inlet. However, because the field radiation level began to increase, the water injection was stopped temporarily. A continuous water injection line was set up between a fire-prevention water pit (water source) and the water supply inlet, whereupon water injection into the reactor using a fire engine and fire-extinguishing line restarted."

Although I do not know the detailed field situation back then, the lineup to inject water into the reactor using a water inlet behind the turbine building protective door and a fire-extinguishing line must have been established. Conversely, radiation could flow out via the pipe from the reactor. Since the reactor (containment vessel) pressure was about 0.8 MPa at around 4 a.m., it is likely that the "field radiation level increase" mentioned in the report was caused by the radiation leaking through the pipe.

I cannot believe that the water injected from the fire engine flowed in the water pipe smoothly while completely filling. Because the containment vessel pressure 0.8 MPa is almost the same as the maximum discharge pressure of the fire engine, I can imagine a flow condition whereby the injected water and gases from the containment vessel pushed each other, which meant the injected water sometimes partially entered the reactor while gases sometimes flowed out of it. If the sealing of the pipe joints was imperfect, radioactive gases are likely to have leaked. This is one possible scenario.

There is another scenario where at least a small amount of injected water entered the reactor through the connected hose. If so, it is quite natural that an oxidation reaction would have occurred with the red hot and mixed molten materials. This reaction may have changed the condition of the reactor and thus the containment vessel, resulting in radiation leaking from the latter. There is strong evidence for this scenario, namely the presence of radioactive tellurium 132 detected in Namie, a town 7 km from the plant at 8:30 a.m. on March 12. Since the half-life of tellurium 132 is 3.2 days, it must have leaked from Unit 1 fuel. Judging from the fact that it reached Namie, its release point is deemed to have been a high stack, not a source on the ground.

Anyway, this event can be interpreted as radioactive leakage from Unit 1, although this conclusion is insufficient to fully convince related parties, including myself, who are pondering and vigorously debating this event. It takes time to con-clude which scenario is correct, so I will refrain from discussing this subject further, by ambiguously describing it as "a condition change due to water injection through fire engine" and move on to the next subject. [Appendix 4.1]

If these study results are correct, it means we have unknowingly ignored a key fact for nuclear safety to date, namely the excellent decontamination efficiency of the SC vent. Its effectiveness is significantly high, and had the Unit 2 vent been opened, the amount of radioactive materials released from the Fukushima accident would have decreased to a few mSv/y, which is lower than the radiation level range for which evacuation is recommended by the ICRP.

Immediately after the vent was opened in Unit 1, the radiation level soared (③), although the amount of released radiation was insufficient to increase the background radiation level. In Unit 3, the vents were opened three times from the morning on March 13 to around noon on March 14, and a massive amount of gases was released (④, ⑤ and ⑥). The rapid decrease in containment vessel pressure (Fig. 2.16) shows the scale of the release. Regardless of this massive release, the background radiation level barely increased and remained almost constant at about 4 μSv/h. This indicates that the amount of radioactive materials released through the SC vent was much smaller than that caused by leakage (①) as a result of the water injection from the fire engine.

The fact that the Unit 3 vent performed three times did not affect the background radiation level caused by the direct release from Unit 1 cannot be reasonably explained without assuming that the radiation released from the vent was lower than the radiation directly released from Unit 1 by at least one digit. Now I am going to discuss the assumption of "one digit smaller." Most of radioactive materials from Units 1 and 3 were released through the vent, excluding those directly released as a result of water injection from a fire engine. The background radiation level due to such radioactive materials remained unchanged and was generally constant at 4 μSv/h until the radioactive release from Unit 2. Based on the aforementioned assumption, namely that the radiation caused by the SC vent was smaller than that caused by direct release by one tenth, the background radiation level resulting from vented radiation was 0.4 μSv/h. Conversely, the background radiation level caused by the radiation from the molten core at Unit 2 was about 300 μSv/h. Comparison of these two cases reveals a radiation removal effect or decontamination factor of SC of about 750.

The aforementioned radiation level 0.4 μSv/h is only dozens of times larger than 0.007 μSv/h, which is the radiation level under normal operation. The SC vent, whose decontamination factor is as high as 750, was quite effective. I estimate that the radiation level of 4 μSv/h, as measured on the morning of March 12, was caused by radiation leaking directly from the reactor core of Unit 1.

I hope future studies will solve this issue in detail. The key point here is the fact that the BWR SC vent system is very effective. Since the currently designed containment vessel SC vent system is effective to such an extent, installing a filter vent would be too redundant and unnecessary. Instead, enhancing the radioactivity removal effect by improving the current design is a more urgent requirement, because unnecessary overlapping of safety systems is sometimes more harmful.

The great effectiveness of the currently designed SC vent systems is very favorable news for nuclear safety. Under future emergencies, SC vents should be opened without much hesitation after isolating the containment vessel, since they are very effective in preventing radiation disasters.

I pity those who made significant efforts at the Prime Minister's official residence to obtain permission to open the vent.

The radiation level increase caused by the explosions at the reactor buildings (Unit 1: between ② and ③ and Unit 3: ⑦) was modest in both cases. If these explosions had damaged the containment vessels, radioactive materials would have directly leaked from the damaged sections, significantly increasing the background radiation level. However, as shown in the Fig. 4.1 radiation level record, no such evidence is suggested, which proves the impermeability of the containment vessels remained intact even after the explosions.

Unit 1, where only the 5th floor was damaged by the explosion and the containment vessel integrity was unaffected, is excluded from this study. However, even the explosions at Unit 3, which ignited on the 5th floor and then propagated to the lower floors, did not increase the radiation level. While the robustness of the containment vessel was proved, I feel it was lucky that it was undamaged.

I said "lucky" because there is a possibility that, at Unit 3 before the explosions, hydrogen was leaking from the first floor equipment transport hatch of the containment vessel (see Part I, Chap. 2, Sect. 2.6). If the hypocenter of the explosions had been the first floor, the impact of the explosions of leaked hydrogen would have deformed the hatch and caused radiation to leak from the containment vessel.

This good luck was attributable to the high hypocenter of the explosions. The first explosion occurred on the 5th floor, which triggered the next explosion on the lower floor and prevented the first floor from becoming the hypocenter of the explosions. If the hypocenter of the explosions had been lower, the radiation level would have been different. I hope future studies will verify further details.

4.1.2 Background Radiation Level Increase for the Second Time – Amount of Radioactive Materials Released from Unit 2

The background radiation level 4 μSv/h changed at around 10 p.m. on March 14 due to radioactive materials being released from Unit 2. Venting was also attempted at Unit 2. The vent did not open, thus, the containment vessel pressure increased and radioactive materials leaked from the molten core directly outside. The concentration and amount of these materials were significant.

It is said that why the vent valve did not open was that the rupture disk inserted in the vent pipe did not break, although opinions seem to differ somewhat. Anyway, the problem with Unit 2 was the vent failure, so I will continue to discuss without going into detail and to proceed assuming the rupture disk failed to break.

A vent during an emergency means a device that releases gases containing radioactive materials from a containment vessel into the atmosphere and a vent also means such a releasing operation. The role of a containment vessel is to confine major reactor systems and equipment in a sealed state, thereby preventing the unplanned release of radioactive materials. To achieve this objective, a pipe that penetrates a containment vessel is equipped with isolation valves before and behind the penetration as a measure to counter pipe rupture and leakage. Isolation valves are also installed before and behind the penetration to prevent leakage from one valve, even when the other valve has failed.

Since a containment vessel is the final protective wall to prevent radioactive materials from being released outside, high requirements are stipulated for its airtightness and its inspections are also strict. A vent, which is normally used only during emergencies, was equipped with not only double isolation valves but also a rupture disk to ensure the prevention of pipe leakage.

A rupture disk is designed to break under a certain pressure and those of Units 1 and 3 broke as designed but that of Unit 2 failed to break. In many cases, a rupture disk fails to break because of its defective installation.

Although hindsight is 20/20, the vent is one of the key safety devices, so its rupture disk should have been designed to be broken by an external force in case it failed to break under certain pressure. This precaution was lacking, which is a design error.

Because the rupture disk failed to break at Unit 2, hydrogen gas generated by the core melt at around 10 p.m. on March 14 increased the containment vessel pressure to about 0.8 MPa. This increase in pressure lifted not only the containment vessel head but also the 600-t concrete shield plug on top of it. The clearances created saw significant hydrogen gas blow out to the refueling floor. At Units 1 and 3, hydrogen gas that leaked in this manner caused explosions.

Conversely at Unit 2, a blowout panel that had been attached to a reactor building wall came away due to the impact of the explosion from Unit 1. Through the opening where the blowout panel had been in place, hydrogen gas flowed outside in a plume (smoke-like assembled gaseous streams) (See Fig. 2.23). Consequently, although no explosion occurred at the Unit 2 reactor building, radioactive materials in the containment vessel were released directly outside with the release of hydrogen gas, increasing the background radiation level and triggering the evacuation of the residents in the vicinity.

At Unit 2, hydrogen gas was also generated thereafter. At around 6 a.m. on March 15, the containment vessel was no longer able to withstand the overpressure and was reported to be damaged, although the specific part remains unknown. For reference, the design pressure of the containment vessel was about 0.4 MPa. This damage rapidly decreased the containment vessel pressure. At around 6 a.m. the pressure descended to 0.2–0.4 MPa (Fig. 2.11). There is no doubt that the containment vessel was damaged.

According to changes in the containment vessel pressure, the situation of radioactive release also changed. The first stage of the release was the leak of hydrogen gas through the opening of the panel (see Fig. 4.1. ⑧). This hydrogen gas was generated as a result of core melt, meaning the concentration of radioactive materials released concomitantly with the hydrogen gas from the molten core was high. Such radioactive materials were released without being decontaminated through the suppression chamber (SC), unlike those from Units 1 and 3. High-concentration radioactive materials were directly released outside, which explains the second background radiation level increase in Fukushima.

Radioactive materials released through the opening of the panel spread and were carried by the wind, which partly reached nearby the main gate. Fig. 4.1 indicates that the measured radiation level temporarily increased to 4,000 µSv/h. This high radiation level decreased a few hours later to 100 µSv/h. However, due to the containment vessel damage that occurred at around 6 a.m. on March 15, the radiation was once again released, and its level recorded a peaking at over 10,000 µSv/h.

This radiation level is very high, given that it is equivalent to an air radiation level of 1 roentgen per hour, in old-fashioned terms. Evacuation of the residents in the vicinity became inevitable. This radiation level increase (⑨), which resembles ⑧, differed from the temporarily spiking radiation pattern previously observed and the radiation release continued for at least a few hours, as shown in Fig. 4.1.

The following summarizes the radiation release from Unit 2: The initial increase in radiation level was caused by significant hydrogen gas generated due to the core melt and released from the opening at around 10 p.m. on March 14. The second increase in the radiation level on the morning of March 15 was caused by radioactive materials released from the damaged containment vessel, which continued for long hours. In both cases, radioactive materials were directly released from the molten core and therefore caused high radiation levels. Even after the containment vessel pressure had decreased to about 0.2 MPa on the morning of March 16, a background radiation level of as high as about 300 µSv/h persisted.

Although it was fortunate for Unit 2 that no explosion occurred there, radioactive materials were directly released from the molten core and triggered serious radioactive contamination in the vicinity. The radioactive disaster that triggered the evacuation of the residents in Fukushima is deemed attributable to the radiation release from Unit 2.

Fig. 4.1 shows the increase in radioactivity that took place at least six times, namely increases ①, ⑧ and ⑨ discussed above as well as increases ② to ⑦ occurring late at night on March 12 until late at night on March 14. These increases in radioactivity were caused by vent operations and explosions as discussed below.

The radiation level depends on meteorological phenomena such as wind speed and direction, weather and humidity as well as accident conditions such as molten core temperature and the presence or absence of the boiling of water accumulating in a containment vessel SC. This complicates the interpretation of radiation level in the first place. However, a change in the radiation level, particularly a rapid increase, suggests that a significant change takes place in the reactor just before or during the change. It is imperative to examine such data with utmost care when clarifying the causes of a nuclear accident.

The following are a review of each item discussed above:

② (Fig. 4.1) indicates an increase in the radiation level due to the temporary vent operation recorded in TEPCO's document.

③ is deemed attributable to radioactive release due to seawater injection at Unit 1.

④ and ⑤ are likely attributable to the vents of Unit 3 reactor.

Although ⑥ is attributed to the ventilation of Unit 3 reactor in the document, a dosimeter indicated a dose increase earlier than the vent operation, the cause of which remains unclear. The release of radioactive materials from Unit 1 can be suspected but details are unknown.

⑦ is an increase in the radiation level due to the explosions at Unit 3 reactor building.

This concludes my study results regarding the background radiation level and its rapid increases during the accident.

4.2   Emergency Evacuation

I think that the background radiation level of 300 µSv/h recorded on March 15 can be used as a maximum reference value for a LWR core melt accidents. If data on meteorological conditions are corrected, this reference value will facilitate future reactor designs. I regard this reference value as the maximum because the measurement point was only 1 km from the molten core and the radioactive materials were directly released from the containment vessel.

Here, why don't you think about the difference between first release through the blowout panel and the second release:

The plume-like radioactive materials ⑧ initially released from the 5th floor into the atmosphere were observed immediately after the core melt. Accordingly, the reactor core had not yet completely melted at the time and most of these radioactive materials are deemed to have been those whose boiling points are low, such as noble gases, iodine and cesium.

The second peak ⑨, which started around 9 a.m. on March 15, was caused by radioactive materials released near the ground from a containment vessel crack that reportedly occurred at 6 a.m. These radioactive materials were slowly diffusing above the ground toward the detector following the air flow while spreading around, and took a certain time to pass away from the detector. Within half a day of the reactor core melting, it is likely that the molten core inside its eggshell like container had come to the boil due to its decay heat. The major radioactive materials released are deemed to have been iodine and cesium, although it is highly likely that various other radioactive nuclides with high boiling points were also mixed in. Their concentrations were probably exceeding those of radioactive materials released in ⑧.

Depending on the wind strength, many of these heavy radioactive nuclides released into the atmosphere probably fell down around the containment vessel during their expansion, while light radioactive nuclides were probably carried far away by the wind. These result in measurement data ⑨.

As I already noted, radioactive contamination depends on wind direction and speed, distance, and other factors. If radioactive materials drifting in the atmosphere unfortunately encounter rain, they fall to the ground and create localized areas of high contamination. A typical example of this is the Gomel region during the Chernobyl accident. Although the region is as far as 120–200 km from the Chernobyl Nuclear Power Plant, radioactive materials of a volume almost equivalent to that observed in the 30 km evacuation area fell to the ground due to rain and created highly contaminated spots in the region (see Fig. 4.2).

Accordingly, since radioactive contamination is highly dominated by meteorological conditions, it is not necessarily appropriate to conclude that the radiation level in the vicinity of the plant main gate indicates the maximum. However, it can generally be assessed that the radiation level in residential areas, which are far from the Fukushima Daiichi Power Station, is lower by about one tenth of those measured at the main gate. This "one tenth" is a rough estimation based on the fact that radioactivity expansion is proportional to the square of distance and assuming the residential areas are about 3 km from the plant.

Look again at Fig. 4.1. The increase in the background radiation level near the main gate, 4 μSv/h at around 4 a.m. on March 12 is equivalent to around 20 mSv in annual dose [Appendix 4.2].

The background level in the residential area is one tenth of the measured site, which is well below the evacuation level of 20–100 mSv recommended to the Japanese Government by the International Commission on the Radiological Protection (ICRP), suggesting no need for emergency evacuation.

Next, the background level, 300 μSv/h, measured after the second increase in radiation level starting at around 10 p.m. on March 14 is equivalent to 1,500 mSv in an annual radiation dose. The annual radiation dose in the residential area is estimated to one tenth of that, namely 150 mSv, which requires emergency evacuation. In other words, according to the recommendation by the ICRP, the evacuation of residents was not necessary until late at night on March 14.

However, the Government suddenly and forcibly implemented evacuation without even prior notice on March 11, the day of the earthquake. It is said that the evacuated residents had no time to prepare and had to ride on buses as ordered without knowing where they were headed. According to the National Diet of Japan Fukushima Nuclear Accident Independent Investigation Commission, among hospital inpatients and people in nursing care facilities, at least 60 people died due to the emergency evacuation. The government, which carried it out forcibly without clear evidence, is highly responsible [1].

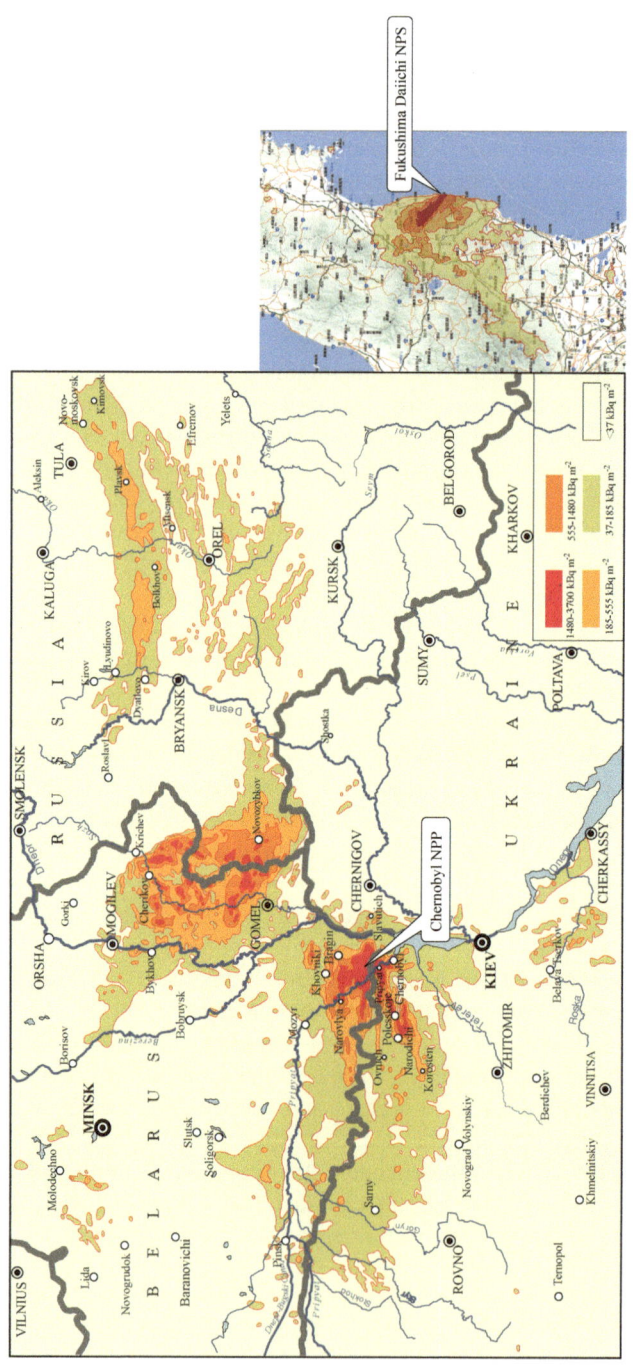

Fig. 4.2 Comparison of contaminated areas between Chernobyl and Fukushima (Contamination situation in Ukraine is based on the data of http://www-pub. iaea.org/mtcd/publications/pdf/pub1239_web.pdf and that in Fukushima is based on the data of http://www.pref.ibaraki.jp/important/20110311eq/20110830_01/ files/20110830_01a.pdf)

4.2.1 Dose Band for Evacuation Recommended by the ICRP

A dose band recommended by the ICRP to the Japanese Government for emergency evacuation is 20–100 mSv/y, with the dose criteria decided in 2007. Because Japan did not legalize the ICRP recommendation on the dose for emergency evacuation, the ICRP suggested its recommendations again to provide support for the Fukushima accident.

Although the recommended dose span is wide, ranging from 20 to 100 mSv, the upper limit 100 mSv is a value the ICRP scientifically judged as not affecting the human body. As for the lower dose limit of 20 mSv, if a nation has come to use this value as its dose limit, this is out of necessity before the ICRP decides its significance. Once the 20 mSv is adopted by the country as its dose limit, it would establish a precedent, in other words, become a social norm in the country. Because it is not easy for the country to revise the standard and related laws once stipulated, the ICRP would recommend wide-ranging dose limits such as 20–100 mSv to save the face of the country. This strategy of offering wide-ranging values is often used in international agreements.

Anyway, the ICRP stated that radiation exposure up to 100 mSv is not harmful to a human body from the radiology perspective, although some scholar groups dispute such view.

The Government of Japan, however, adopted the lower limit of 20 mSv as its dose limit from the beginning. The reason for this option is unclear. Even more incomprehensible is the fact that this government decision was made long after the emergency evacuation had ended. In other words, the Government forcibly performed the evacuation without setting a standard dose for evacuation.

If the Government had performed the evacuation assuming 20 mSv as the minimum dose requiring emergency evacuation, this alone would have changed the situation in Fukushima. This is because, had a stipulated evacuation dose level been established in advance, evacuation would not have been performed until that dose level was reached. Consequently, the evacuation should have been postponed until late at night on March 14 when the annual dose in the residential areas was about 2 mSv/y. Three days had passed after the accident occurred by that time, that meaning the evacuees would have been well prepared mentally and had adequate time to pack the necessary items. The government could have had sufficient time to develop an evacuation plan, which would have prevented traffic jams during evacuation and the tragedies of elderly people dying of fatigue. If the areas to be evacuated had been limited to those reaching the evacuation dose, the evacuated areas would have been far fewer and narrower.

Moreover, initially setting a high evacuation dose and lowering it gradually while observing the situation provides the general public with a sense of safety. Once a low evacuation dose is set, revising it upward makes the public feel insecure. Initial value setting is a key for emergencies. In the case of Fukushima, the evacuation dose should have been set at 100 mSv initially and lowered gradually. If so, the number of people who required evacuation would have been much lower. I think the 20 mSv option adopted by the Government was a mistake.

What I consider more deplorable is the fact that this 20 mSv standard was further lowered and is now set at 1 mSv practically. Gohshi Hosono, the former Minister of State for the Nuclear Power Policy and Administration, promised to set the decontamination level at 1 mSv in a meeting with Fukushima prefectural governor, which practically set the mental permissible level for the evacuees to return homes.

This situation is attributable to mishandling of the emergency by the former prime minister Mr. Kan administration, nearly 3 years after the event. It is high time for cool-headed and scientific judgment.

4.2.2 Lives of Evacuees

These days in Japan, emergency evacuation is positively perceived as if it is a humanitarian measure, since it is considered to be a quick removal of residents from risks.

However, while it takes only one day for evacuation, stays in shelters are long. In shelters, people undergo long hopeless lives without clear outlook. During the Second World War, I was separated from my parents and experienced virtually compulsory schoolchildren evacuation, so I know how hard and tough the evacuation life is.

Evacuees in Miyagi and Iwate prefectures, who suffered the tsunami disaster due to the Great East Japan Earthquake like those in Fukushima prefecture, have already returned home and are striving to realize the dream of reconstructing their homelands. Their faces are bright with smiles.

In a small restaurant in Ishinomaki, Miyagi pref. in September, half a year after the earthquake, I saw a poem which expresses the joy of having survived the disaster honestly and frankly. Touched by this poem, I am impressed that it expresses strength and hope for an energetic and bright future so well. I also felt the same way when I was able to return home from wartime evacuation.

I am afraid to say, but in my eyes, the mindset has become quite different between the people of Ishinomaki and those of Fukushima prefectures, although both peoples suffered the same Great East Japan Earthquake.

Evacuees in Fukushima are said to be divided into those who want to return home and those who do not. This is probably because they were forcibly evacuated because of the nuclear accident in addition to the earthquake and tsunami. Moreover, there are probably many helpless people who want to return home and reconstruct their lives but cannot do so, partly because they are bound by the Government's promise to decontaminate their hometowns to 1 mSv. It is quite understandable that such helpless feelings turn to hatred against the nuclear industry and TEPCO.

However, if the reason for their lack of hope is the fact that they cannot return home or hesitate to do so because of the radiation level, calm study of radiation would help.

I have worked in nuclear sites for about 60 years and was engaged in various types of work under radiation. I have experienced not only plant operation and maintenance work but also decommissioning as well as experiments on fuel rod melting and damage in safety studies. I also entered the sarcophagus (stone coffin) of the Chernobyl Nuclear Power Plant to observe the post-accident situation in details. I have been exposed to radiation and ingested more radioactive materials than other people. Despite reaching 80 this year, I have no problem with my health.

Based on such experiences, I state my personal views. I do not think that the radiation level in Fukushima is harmful to human bodies except for the high-radiation area in the plant site vicinity. Those who want to return home should do so as soon as possible.

The accident took place, undeniably, but the future should be constructed with hope.

I believe that calm understanding of the radiation effect on health and acting with hope for future will promote the reconstruction of Fukushima and people's health; both physically and mentally.

4.3   Release of Radioactive Materials and Contamination

So, how wide was the radioactive contamination spread by the Fukushima accident?

Fig. 4.2 shows a comparison of the contaminated areas between the Chernobyl and Fukushima accidents using the same gauge. In fact, the contaminated areas in Chernobyl are wider than those of Fukushima; even too wide to be covered in this page. In addition, the contamination map for Chernobyl shows a distribution 5 years after its accident whereas that for Fukushima shows 1 year after its accident. Although a direct comparison is not meaningful, it will help in roughly understanding of the radioactive contamination area. How the expansion of contaminated areas differs between the two cases is clear at a glance.

Next, the amount of radiation due to the Chernobyl accident was about seven times larger than that of the Fukushima accident. Moreover, the power of the one molten reactor in the Chernobyl NPP was about 1,000 MW while the combined power of the three molten reactors in the Fukushima Daiichi NPS was about double at 2,000 MW. This means that the radioactive release per reactor power of the Fukushima Daiichi NPS was about one 15th of that of the Chernobyl NPP.

The reasons for this significant difference include natural conditions such as mountainous geographic features in Japan and wind directions during each of the two accidents but another major reason is the difference in reactor type. The fatal causes of the spread of radioactive materials over such a large area as happened at the Chernobyl accident were that a fire broke out, which lifted radioactive materials high into the air and the reactor in question had no containment vessel to confine radioactive materials. The following describes the situation of radioactive materials released in Chernobyl, although it may deviate from the main theme of this book. Please compare it with the current situation in Fukushima.

4.3.1 Radioactive Materials Released from the Chernobyl Accident

The Chernobyl NPP's reactor core is a graphite block of about 11.8 m in diameter and about 7 m high, which has about 1,700 holes of about 10 cm in diameter each, each housing a cooling water tube made of zircaloy (standard nomenclature: zirconium-Niobium alloy) containing 18 fuel rods. These are called core internals (Fig. 4.3). The reactor core is surrounded by a thin iron plate, which is further surrounded by a thick concrete cylinder. This concrete cylinder plays the role of an LWR reactor vessel.

During the Chernobyl accident, explosions and a fire took place in the reactor. The details are omitted here because of their long process [2].

It would be easy to understand the Chernobyl accident by recalling briquette coal, which was used for space heating in every household in Japan until 30 years ago. As I remember, the briquette coal is a cylindrical block 15 cm in diameter and height, made of solidified coal powder and with 12 vertical through-holes 1 cm in diameter. Except for their scale, the briquette coal and the Chernobyl reactor core look quite look alike in terms of morphology. Moreover, because the graphite used

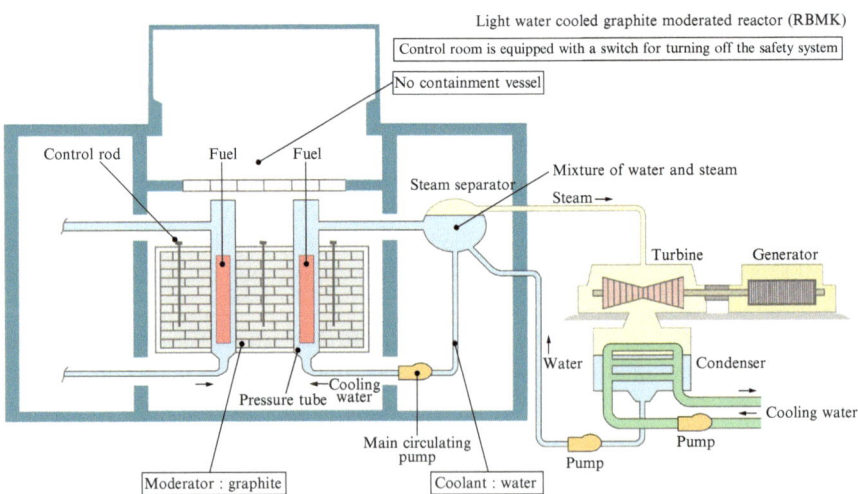

Fig. 4.3 Structure of the Chernobyl NPP (Source: a pamphlet from the Agency for Natural Resources and Energy)

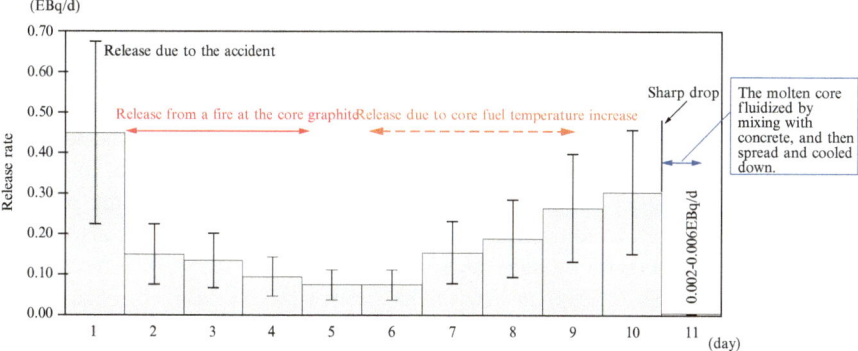

Fig. 4.4 Daily release rate to the atmosphere of radioactive material during the Chernobyl accident (Source: based on a report by the IAEA's Chernobyl Environmental Expert Group)

as the core material is highly pure carbon, the Chernobyl reactor core is almost the same as the briquette coal in material terms. In other words, the Chernobyl reactor is a large coal briquette.

The radioactive release from the Chernobyl reactor continued for 10 days. As shown in Fig. 4.4, after a relatively significant radioactive material release on the first day, the release decreased temporarily, but began to increase again on around the 7th day, a quite strange release pattern. This is because the release of radioactive materials can be divided into three stages, namely a significant release upon accident occurred, release due to a fire at the reactor core, and release from molten fuel.

The Chernobyl accident started with a reactor power excursion and explosions, and on the next day the core graphite caught fire. I still remember the reactor core burning red in a photo taken by the U.S. satellite.

The release of radioactive materials immediately after the accident was caused by a reactivity accident where extremely large heat generated instantaneously, and resulted in not only the melting of fuel uranium dioxide but also the release of partially vaporized radioactive materials into the air. Radioactive materials contained in the fuel rods were released because of their damage. Fuel rods scattered outside the reactor by explosions also spread radioactive materials. The amount of radioactive materials was enormous, including those with a short half-life released immediately after the onset of the accident.

Although the main radioactive nuclides initially released were gaseous radioactive materials such as noble gases accumulating in the fuel gap, radioactive materials with a melting point of about 3,000 °C such as plutonium were also released. This is the major characteristic of a reactivity accident. However, the radioactive contamination with high melting point materials due to such radioactive release was limited to the area in the reactor vicinity.

From the second day, vaporized and gaseous radioactive materials were released due to the fire in the graphite constituting the reactor core and spread all over the Northern Hemisphere accompanied by fire smoke.

It is said that, on the morning of the first day of the accident, about 5 h after its occurrence, children in Pripyat, a city 5 km from the Chernobyl NPP, saw wispy smoke rising into the air above the NPP. This smoke was a trace of a fire having occurred at the turbine building, caused by fuel rods that had been scattered by the explosions. At the time, the core graphite had just caught fire, but was too small to be called a graphite fire – resembling the burning of ignition material to kindle the briquette coal. Accordingly, there was no problem, even for children playing outside. The amount of radioactive materials that reached Pripyat was also small and most people in Pripyat thought the accident had ended.

That night, people in Priptyat saw the sky above the NPP tinged pink, which made them wonder. At around noon the next day, the radiation level soared, triggering an emergency evacuation.

It is said that, before the briquette coal catches fire, air heated by burning ignition material traverses the through-holes of the briquette coal and raises the temperature of its entire body to about 700 °C, triggering spontaneous combustion In the same process, after the Chernobyl reactor core graphite started spontaneously combusting, the number of upward currents ascending through the 1,700 holes increased and the hot upward currents expanded in the distance. This is why the radioactivity level in Priptyat soared, triggering the emergency evacuation. The pink light that tinged the night sky on April 26 was probably from the flames of the reactor core fire.

The temperature of the graphite constituting the Chernobyl reactor core reportedly reached at least 1,200 °C and some say it reached 1,500 °C. Hot carbon dioxide plumes heated by the large-scale briquette coal rose into the sky through the ceiling destroyed by explosions.

The radioactive materials released from Chernobyl were carried by jet streams and reached various locations in the Northern Hemisphere. Unless I am mistaken, they reached Hokkaido 5 days later. Radioactive materials blown up high in fire plumes were carried by jet streams flowing in the sky at altitudes of 5,000–10,000 m before reaching various locations of the Northern Hemisphere. These were huge plumes beyond my imagination.

This reactor core fire made radioactive materials with a boiling point of less than 1,200 °C to vaporize and scatter with fire plumes around the world. Conversely, radioactive materials with a boiling point exceeding 1,200 °C were cooled by the fire and preserved in pellets. Although "cooled by the fire" sounds ridiculous and nonsense, this actually happened. For fuel pellets (uranium dioxide) with a melting point of 2,880 °C, a 1,200 °C fire is as low as cold water. During the reactor core fire at the Chernobyl NPP, fuel rods were cooled by the CO_2 flow generated from the fire. Like a briquette burns to ashes from the top down, the graphite burned and sublimed from its top downward, which meant the reactor core shortened over time and gradually decreased the radioactive material release.

This is proved by the fact that the gigantic reactor shield plug that dropped and pierced the core graphite obliquely after the explosion increased in obliquity and stood almost vertically in the end. The shield plug had somersaulted in the air due to the explosion and dropped onto the core graphite. Because the core graphite onto which the shield plug had landed shortened as the graphite burned, the landing point (point of load application) also descended. Consequently, the shield plug stood almost vertically (Fig. 2.18).

The decrease in radioactive release shown in Fig. 4.4 resulted from a decrease in the core graphite due to its burning. The amount of released radioactive materials also decreased on the 5th day until the next day (April 30–May 1), which was the period when the graphite fire ended. The core graphite burned (sublimed) completely. At this point, air flow caused by the fire stopped. Inside the gigantic cylindrical concrete shield wall surrounding the reactor core, air flow that had been cooling the fuel rods also stopped. It was natural that the temperature of the fuel rods having decay heat would increase and this temperature increase triggered the third release of radioactive materials.

The length of the Chernobyl fuel was about 7 m. After the graphite had burned and disappeared, there was nothing to support the lanky fuel rods and they inevitably collapsed, probably during the fire.

It is likely that the fuel rods buckled or leaned against each other and finally came to rest on the cylindrical concrete shield floor. Once these collapsed fuel rods were deprived of fire-caused heat dissipation, their temperature soared and finally melted. When their temperature exceeded 1,200 °C, which was the fire temperature, the radioactive release restarted. It is likely that the radioactive materials with high boiling points preserved inside the fuel rods vaporized sequentially, exited the cylindrical concrete shield wall, were cooled by the outside air, solidified and dropped onto the ground around the reactor. Some may also have been blown away to distant places. This was the situation of radiation release on the sixth day onward. The fact that the release increased on and after the sixth day indicates that the temperature of the collapsed fuel rods increased.

However, this release stopped suddenly on the tenth day. The record simply states "sharp drop" without any explanation. The fuel rod temperature probably reached 2,000 and a few hundred degrees, which is the melting point generating a molten mixture, and the fuel rods started to melt and liquefy. After the solid fuel rods reached their melting point, they required significant latent heat to melt and liquefy. Because their temperature was constant during this period, the radioactive release stopped temporarily, which probably explains the "sharp drop" in temperature on and after the tenth day. In addition, the liquefied molten mixture probably lowered its melting point by melting the concrete shield floor below. About a quarter of the disc-shaped lower shield board (shield plug) of the Chernobyl reactor completely melted down.

As I already noted, the radiation release in the Chernobyl NPP can be divided into three stages, namely the significant release immediately after the accident occurred, the second release in fire plumes scattering radioactive materials widely around the world, and the third release of radioactive materials with high boiling points. The reason why these release patterns were so distinguishable is that the Chernobyl reactor lacked a containment vessel to confine the radioactive materials.

4.3.2   Release of Radioactive Materials from the Fukushima Accident

The release of radioactive materials from the Fukushima accident started with radioactive leakage following water injection through a fire engine pump to cool Unit 1 performed at 4 a.m. March 12, 1 day after the accident had occurred. The radioactive leakage was relatively small; only increasing the background radiation level to about 4 μSv/h. Subsequently, the vent valves of Units 1 and 3 opened and the molten core radioactive materials accumulated in the containment vessels were released through stacks, but this background radiation level remained almost unchanged. Late at night on March 14, however, from the containment vessel of Unit 2, where the SC vent had failed, radioactive materials in the molten core were directly released, elevating the background radiation level to about 300 μSv/h. This radiation level decreased on and after March 17 and kept decreasing after a temporary power supply had been installed on March 20.

Radioactive materials released during this period can be divided into those directly released from the containment vessel and those released through stacks after the SC vent. The radioactive concentration of the former was high and twice increased the background radiation level. Conversely, the concentration of radioactive materials released through the SC vent was diluted by the decontamination effect of the SC water.

A significant amount of radiation from the former, the radioactive materials directly released from the containment vessel, is likely to have been continuously released for about the first 10 days of the accident, during which the containment vessel pressure was high. After a temporary power supply had been installed and core cooling started, gaseous radioactive materials leaking from the molten core probably decreased rapidly, because they were cooled by the injected water, reverted to a solid or liquid state, and mixed into the containment vessel SC water.

Except for noble gases, which were released immediately after the accident occurred, most of the radioactive nuclides released on after the tenth day were iodine and cesium. Because other radioactive nuclides have high boiling points, they were cooled and solidified due to water injected to cool the reactor core, and thus their release was very limited.

The radioactive release decreased rapidly as the containment vessel water temperature (SC water temperature) decreased to less than 100 °C thanks to cleanup & cooling devices started to operate in June (see Fig. 4.5). The amount of radioactive materials released in November 2012, 18 months after the accident, was about 0.01 billion Bq/h, which is about one hundred millionth of the maximum release of 800 trillion Bq/h observed on March 15, 2011 immediately after the accident occurred.

The amount of radioactive materials released from the Fukushima accident was about one 7th of that released from the Chernobyl accident in a rough comparison. In the case of Chernobyl, radioactive materials were directly released outside due to a fire and widely contaminated the Northern Hemisphere, whereas in the case of Fukushima, most of the radioactive materials released immediately after the accident

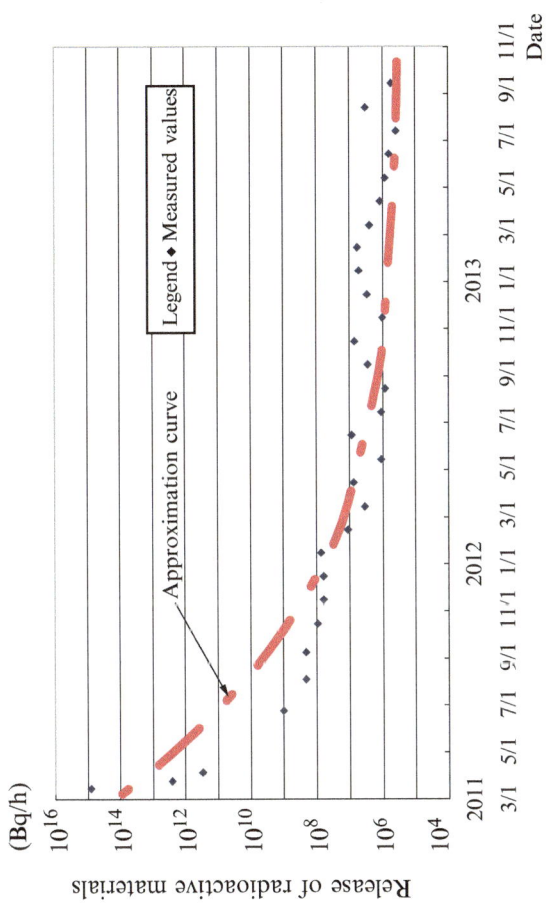

Fig. 4.5 Release rate of radioactive materials in 2 years after the Fukushima Daiichi accident (Source: based on the TEPCO's document)

were iodine and cesium. In the case of Fukushima, other radioactive materials were solidified by core cooling and washed and cleaned through the SC vent, limiting their release. What made this difference was the presence and absence of a containment vessel.

Had it not been for the containment vessel damage at Unit 2, the radioactive release from the Fukushima Daiichi NPS would have been much smaller, probably about one 500th of the release from Chernobyl NPP.

Conversely, the light water reactor, which is equipped with a containment vessel and uses water which prevents fire, should have been safer. Their safety was spoiled because of vent failure.

It is said that the widespread contamination in the Gomel region east of the Chernobyl NPP and the contamination that spilled further eastward to Minsk from the Republic of Belarus, were caused by radioactive materials being blown up into the sky by the reactor fire and falling down with rain. Although not shown in the map, many cases of contamination were reported such as in the northern Finland tundra area and pastures in the Dolomites region of the Italian Alps.

Because the reactors of the Fukushima NPS are LWRs, a fire that blows radioactive materials high into the sky cannot occur. Accordingly, it is likely that radioactive materials were carried by land winds blowing at relatively low altitudes and fell onto the ground, and their contamination was somewhat blocked by mountain slopes and trees. Although it may not be precise, I heard that the wind directions within a few days of the accident were generally seaward, while on March 16, wind started to blow northeast toward the Iitate-mura.

Table 4.1 shows a comparison of the radiation exposures and evacuation situations in Fukushima and Chernobyl as well as the TMI where the core melt also occurred.

Table 4.1 Comparison of disasters among the TMI, Chernobyl and Fukushima accidents

| | Number of evacuees | Exposure (mSv) | | Source |
		Maximum	Average	
TMI	About 1,000 (voluntary evacuation: 150,000)	About 1 mSv	About 0.01 mSv	*Reactor Power Excursion* p. 361 [1]
Chernobyl	About 130,000	About 5,000 mSv	About 100 mSv	
Fukushima	About 157,000 (1)	25 mSv (2)	0.8 mSv (2)	(1) Total number of evacuees due to earthquake, tsunami and nuclear accident (announced by Fukushima prefecture's Emergency Response Headquarters on December 6, 2012) (2) Fukushima pref. 13th Health Care Research Committee's document (November 12, 2013)

Radiation exposure caused by the TMI accident, where the molten core was cooled within a short time and the containment vessel integrity remained intact, was small in terms of the number of exposed persons and radiation level. The number of people evacuated forcibly was about 1,000, although their average exposure level was small at 0.01 mSv. However, according to an unofficial report, the number of people who voluntarily evacuated in compliance with evacuation instructions exceeded 150,000.

Conversely in Chernobyl, 29 firefighters who worked on site to extinguish a turbine building fire immediately after the accident were exposed to a significant amount of radiation and died, and two plant workers died of explosion. The number of people forcibly evacuated was 130,000, whose average exposure level was about 100 mSv.

Conversely in Fukushima, the number of evacuees is estimated at about 160,000 and their average exposure level about 1 mSv (both according to research by Fukushima prefecture).

It should be noted by comparing the three accidents that, while it may be a coincidence that the total number of evacuees was relatively similar, the average exposure levels are shown in a geometrical series. I reaffirm that if only the vent of Unit 2 had not failed, the average exposure level would inevitably have resembled that caused by the TMI accident.

Table 4.1 shows that the disaster caused by the Fukushima accident was less serious compared to the Chernobyl accident, regardless of the fact that the former triggered core melt or explosions in as many as four units. Probably because of this fact, there is international opinion that the International Nuclear Event Scale (INES) concerning the Fukushima accident should be revised to Level 6, not the current Level 7, which is equivalent to that of the Chernobyl accident. I consider this opinion quite valid.

Related parties in the U.S., England and France versed in nuclear power generation have frankly expressed surprise at the fact that, following the large-scale natural disaster, no-one died due to the Fukushima nuclear accident and radiation exposure was not serious.

A good example is the comment of prominent environment journalist George Monbiot posted on Britain's Guardian on March 21, 2011 immediately after the accident took place. He noted regarding the Fukushima Daiichi NPS accident that "Despite the natural disasters where tens of thousands of people died, no-one died of radiation due to the nuclear accident. Due to the disaster at Fukushima, I am no longer nuclear-neutral. I now support the technology… Yet, as far as we know, no-one has yet received a lethal dose of radiation." I am also of the same opinion.

Although the Nuclear Regulation Authority of Japan has revised and is enforcing the Safety Design Standard, strong concerns about the revision have been expressed, particularly by some European countries and the U.S. For example, while there are moves to strengthen the IAEA's international nuclear safety standards, there is no movement for revising it. Frankly speaking, despite the national sentiment in Japan, opinions in European countries and the U.S. diametrically opposed to those in Japan are due to the mildness of the disaster caused by the Fukushima Daiichi NPS accident as shown in Table 4.1. I guess that leaders in those countries have renewed their confidence in the strong safety of light water reactors, even though they do not publicly say so. Accordingly, many people worldwide are requesting the Japanese Government to restart nuclear power plants as early as possible.

Now I summarize this chapter. Changes in the background radiation level due to the release of radioactive materials from the Fukushima accident started at 4 a.m. on March 12 with a small amount of ground-level release following water injection into Unit 1 from the fire engine there. The background radiation level further increased due to the ground-level radioactive release from the molten core in Unit 2 at 10 p.m. on March 14. The former release lasted for about 3 days from March 12–14 and caused a radiation level of 4 μSv/h. The latter release dominated the subsequent radiation level over the plant-surrounding area, reaching a radiation level of 300 μSv/h.

Had the vent operation at Unit 2 been successful, thereby decontaminating the radioactive materials through the SC, the radiation level would have been limited to a few μSv/h as in the case of Units 1 and 3. If so, almost no areas would have had background radiation levels exceeding the minimum value of 20 μSv/y recommended by the ICRP and there would have been no evacuation.

They say that many evacuees are irritated by their long shelter lives, uncertain when they can go home and have lost hope for the future. While there may be various reasons why they cannot go home, one of which is the nonsensically low annual radiation level of 1 mSv, which is hampering their return home like an evil spell. Politicians must correct this distorted situation so that those who want to return home can do so.

The good news is the fact that the SC vent has great decontamination efficiency, although this often goes unnoticed in the shade of the direct radioactive release from Unit 2. If it had operated perfectly, it would have been possible to limit the radiation level under the accident to only dozens of times of that during normal operation. While I will leave this issue to future studies and verification, I believe it is not necessary to install a filter vent.

Appendices

Appendix 4.1

On December 13, 2013, I was finalizing this book, when TEPCO newly reported on the seawater injection situation. It said that the pipe that had been used to inject seawater into the reactor from the fire engine had several branch pipes, some of which had outlets at the turbine building. If so, the sealing of hose-connecting joints is no longer an issue. Radioactive materials must have leaked from the reactor into the turbine building through the branch pipes of the pipe used to inject seawater into the reactor. Although related parties, including myself, had vehemently insisted on either of the two views, both proved correct. This report by TEPCO was very helpful for me in writing this book. Therefore, TEPCO's announcement did not affect my study results, but rather corroborates them.

Appendix 4.2

Radiation Level and Annual Exposure Dose

One day is 24 h and 1 year is 365 days. When converting the radiation level to an annual exposure dose, the former is multiplied by 365, assuming ① a person spends 8 h a day outdoors and the remaining 16 h indoors and ② the indoor radiation level is four tenths of the outdoor radiation level.

For example, if the radiation level is 1 μSv/h, the annual exposure dose is 5 mSv.

References

1. Ishikawa M (1996) Reactor power excursion. Nikkan Kogyo Shimbun, Ltd. [Japanese version only]
2. The official report of The Fukushima nuclear accident independent investigation commission. Page 30 of Chapter 4, July 2012. http://warp.da.ndl.go.jp/info:ndljp/pid/3856371/naiic.go.jp/en/

Chapter 5
Tsunami and Station Blackout

The Fukushima Daiichi NPS accident was triggered by a blackout (loss of external power supply) due to the earthquake and by the lack of any electrically operated emergency safety systems and equipment because emergency power supplies, distribution panels and switchboards were flooded by the tsunami. In addition, recovery from the blackout, on which the electric power company had been confident not to take more than eight hours, was actually very slow and only realized 10 days later when a temporary power supply was installed. Accordingly, although the core cooling system of each of the Units 2 and 3 operated without electricity, better than designed, the core melts occurred at all of the Units 1–3.

Here, I discuss why the core melts and explosions at Units 1–3 did not take place at the same time, even though they were attacked by a simultaneous earthquake and tsunami (see Table 5.1).

The hydrogen explosion in Unit 1 occurred at around 3:30 p.m. on March 12, the core melt at Unit 2 at around 10 a.m. on March 14, and the hydrogen explosion at Unit 3 at around 11 a.m. the same day. There was a significant time difference exceeding 2 days with regard to the core melts and explosions among the three units, which was due to the variation in the performance of non-electric safety devices at each unit. The differences in their operating time period resulted in a difference in the timing of core melts and explosions.

Although mass media reports suggested the reactor safety equipment had been totally powerless due to the blackout, devices for cooling the reactors without electricity operated to prevent a core melt for some time period. Although the IC in Unit 1 failed to operate, the RCICs in Units 2 and 3 performed much better than designed. Thus the NPP's existing safety devices were far from useless. If electricity had been restored during the operation of these safety devices, the Fukushima accident would have been very different.

In this regard, the slow power recovery was one of the factors that exacerbated the accident into a disaster. This was the second factor of the accident. In this chapter, I examine the station blackout and the tsunami that triggered it.

© Springer Japan 2015
M. Ishikawa, *A Study of the Fukushima Daiichi Nuclear Accident Process*,
DOI 10.1007/978-4-431-55543-8_5

Table 5.1 Core melts and hydrogen gas explosions in the Fukushima accident

Date	Unit 1	Unit 2	Unit 3	Unit 4
March 11	Earthquake at 2:46 p.m.			
	Tsunami at 3:35 p.m.			
March 12	Core melt at around 4 a.m.			
	Hydrogen explosion at 3:36 p.m.			
March 13			Core collapse at around 9:30 a.m.	
March 14			Core melt at around 10 a.m.	
			Hydrogen explosion at 11:01 a.m.	
		Core melt at around 10 p.m.		
March 15				Hydrogen explosion at around 6:14 a.m.
Remark		No hydrogen explosion occurred because the blowout panel was open		Shutdown for a periodic inspection at the time of the earthquake

5.1 Seawall (Tide Embankment)

It was hard to believe when I heard the evening news on the day the earthquake occurred on March 11, 2011 that the Fukushima Daiichi NPS had been attacked by a tsunami and damaged. Until then, I believed seismologists that said that NPP sites were high enough to be unaffected by tsunamis.

The following is my experience of 40 years ago, in the early 1970s. I was assigned as a safety review assistant of the Nuclear Safety Commission. Back then I was young, without adequate knowledge other than in my field and studying as a trainee. I remember asking a question regarding a tsunami, which was outside my field, during a safety review meeting. I asked whether the heights of reactor sites were adequate against tsunami. In response, an old member in charge of a geology and seismology subcommittee answered generally as follows:

> "You have probably asked this question because you still remember the story of *Inamura no hi* (a documentary concerning the 1854 Nankai Earthquake) you learned in primary school. That kind of tsunami, which ran up onshore, occurs at a rias coastline with a complicated shape. Tsunamis that attack an open coast along an expansive ocean are big swelling waves with a long wavelength. In this sense, it is an expansive and high tidal wave. In Japan, assuming a tsunami height of no more than 6 m at most is enough."

These words still remain in my memory.

After seeing that tsunami in Fukushima, I read *Tsunami* authored by Akira Yoshimura, a novelist with an acute insight into the progress and development of scientific technologies. His words are reliable and not exaggerated. His book shows the record of a large-scale tsunami, 85 m high, which attacked Ishigaki island in Okinawa in the Pacific Ocean in 1771.

In July 2013, I visited the site to see traces of the tsunami. I found an impressive memorial monument standing on a hilly upland about 100 m above sea level, located in the southeast of the island, which was probably a site victimized by that large tsunami. The monument said about 1,000 people died, which was about half the island's population at the time. The magnitude of the Yaeyama Earthquake, which triggered the tsunami, was 7.4.

The upland commands an unhindered panoramic view of the landscape below, through almost 180° and centering on the southeast of the island. All I could see beyond the slope were white-crested waves beating against coral reefs in the distance and blue sea. It was far from a rias coastline, where tsunamis tend to run up. Although the old subcommittee member's theory that "tsunamis that attack open coasts along expansive ocean are expansive and high tidal waves" may reflect the general characteristics of tsunamis, his theory does not seem applicable to either the Yaeyama tsunami, or to that which attacked the Fukushima site.

Tsunamis do not seem a natural phenomenon that can be defined simply as "expansive and high tidal waves." I think that their scale depends not only on coastal topography but also on seabed topography and the number of earthquakes causing them, and therefore cannot be defined simply. Tsunamis is quite a complicated phenomenon to be studied with the Earth mechanism taken into account. I am unsure as to the progress of tsunami-related academic knowledge over the past 40 years. However, as far as I have observed questions and answers during committee meetings for revising the Regulatory Guide for Seismic Design of Nuclear Power Plants, tsunami-related knowledge seems almost unchanged and on an approximate level.

Despite such level, the then Prime Minister Naoto Kan requested the Hamaoka NPS of the Chubu Electric Power Co., Inc., whose high risk of a large-scale earthquake has been highlighted for 40 years because it is located above the possible epicenter of an anticipated Tokai earthquake, to construct a seawall. It was difficult for the Chubu Electric Power Co. to decline such request and the company is now building a seawall 22 m high, which is said to cost more than 100 billion yen.

Other electric companies, which had been closely watching such progresses, also began constructing seawalls. Because they decided to do so voluntarily after estimating the maximum possible tsunami height, third parties like myself should probably not interfere, but I want to point out one thing. I wonder whether they simulated the Yaeyama tsunami with a height of 85 m using a computer during the process of estimating tsunami height. If not, the scientific validity of the seawalls under construction is dubious, meaning they are wasting their money on meaningless constructions.

I do not oppose the construction of necessary seawalls. For example, a very useful seawall was successfully constructed in Amsterdam, Holland. I oppose, however, the construction of seawalls whose scientific validity is dubious around NPPs. This also applies to regulations. Dubious construction and regulations undermine nuclear safety.

I will explain why I oppose the construction of seawalls. A seawall can block tsunamis only by its height. However, a seawall 22 m high cannot block a tsunami of the same height as the 85-m Yaeyama tsunami. Akira Yoshimura's *Tsunami* says there are traces of a tsunami whose height exceeded 500 m in Alaska. It would be impossible to block such a gigantic tsunami with a seawall. So how would it be harmful?

Suppose a tsunami exceeding 22 m high were to attack the Hamaoka NPS. The expansive and high tidal wave would cover the entire plant and the seawater would fill the inside of the seawall. At this point, the seawall would negatively function as a reservoir for the seawater. Without a seawall, it would be still possible to manage the situation until the tsunami subsides. However, once a site surrounded by a sea-wall becomes a puddle, the water would not be easy to remove, even via check valves for drainage, expanding the disaster. Electric systems, equipment and rotary machines would all be rendered useless.

Moreover, plant workers might drown or suffocate. In the meantime, decay heat in the reactor remains and keeps generating heat. Who would be able to handle such circumstances and how? In this situation, operators like those who bravely fought the Fukushima disaster would not be available. Even if electricity were available, no machine would be operable. A site buried under sludge caused by a tsunami would create an accident environment more severe than the Fukushima Daiichi NPS. It is apparent that the resultant disaster would be not so much a repeat of the Fukushima accident but a more severe disaster. This is the negative effect of a seawall, triggered by a safety structure constructed based on unscientific and ambiguous validity.

Any conventional safety devices manufactured to prevent anticipated failures have had their effectiveness verified based on scientific knowledge and in some cases, were examined for validity, even by the destructive reactor testing. They have been used only after strict examinations to ensure their safety. Despite this, Fukushima failures took place. The reason for the failures of the safety devices established with such careful considerations was that the accident events exceeded the design assumption. The construction of seawalls, as proposed by former Prime Minister Kan on the spur of the moment, seem to be perpetuating the same mistake.

5.2 Station Blackout

The principal cause of the station blackout was the tsunami. The blackout lasted for an extended period of time. Some says that power distribution panels were sub-merged under water and therefore were not connectable or usable even after elec-tricity became available.

What they say is true, albeit not a valid reason why power supply could not be restored. This is because if a small amount of electricity had been available, operators could have taken certain measures. Electric power of 1 kW corresponds to 20 manpower in terms of calorific value and restoring electricity is equivalent to the support of tens of thousands of workers. I think that if the leaders of the Government and TEPCO had immediately made all-out efforts to provide emergency support to recover the power supply by providing temporary power-supply devices and power distribution panels, the expansion of the accident could have been prevented and the disaster requiring the residents' evacuation would not have occurred.

However, public opinions, misled by the mass media, were directed against the former Regulatory Guide for Reviewing Design of Light Water Nuclear Power Reactor Facilities (hereinafter referred to as the former Regulatory Guide), which assumed the duration of station blackout to be "short," but not against the Government's poor emergency support. Consequently, a trend has emerged whereby the safety concept established through international studies and examinations is modified in a makeshift manner. Prompted by general public opinion, the Nuclear Regulation Authority has stipulated the acceptable maximum duration of incidents involving the loss of external power supply as 7 days and simultaneously demanded the 7-day operation performance of an emergency AC power supply. However, I wonder whether this measure will suffice to ensure the power supply.

A station blackout means a combination of incidents involving the loss of the external power supply (blackout in general terms) and the lack of any on-site power-generating equipment and emergency generators, which is not a simple blackout.

Incidentally, the former Regulatory Guide required "the safe shutdown of reactor and post-shutdown cooling in the event of short-period station blackout (gist)" in Guide 9 - Design Considerations against the Loss of Power Supply (June 1977) and stated "It is unnecessary to assume a prolonged power supply loss because the restoration of transmission grid or the repair of emergency diesel generators is expected (gist)."

As far as the Fukushima accident is concerned, the former Regulatory Guide failed. It naturally failed because the duration of both station blackout and the loss of external power supply exceeded the required duration. The statement "the repair of emergency diesel generators" in the former Regulatory Guide was impossible under the tsunami disaster. Under such circumstances, the Regulatory Guide had assumed "the restoration of the transmission grid" as a last resort, which also was impossible to implement, which meant the Regulatory Guide failed to function.

However, had the Government responded calmly, early recovery from the blackout as required by the former Regulatory Guide would have been socially and technically possible. Delaying the recovery from the blackout was fatal. What violated the former Regulatory Guide was the Government, which was unable to utilize national power and technologies nor to take measures.

What was the basis of the provision concerning station blackout "It is unnecessary to assume a prolonged power supply loss" stated in the former regulatory guide? There were two reasons for this provision. One was the track record of the duration and frequency of blackouts in the U.S. (Table 5.2) and the track record of blackouts in Japan (Table 5.3) at the time when the former Regulatory Guide was under development. The other was the presence of safety equipment capable of cooling a reactor, even without electricity.

Table 5.2 The track record of the duration and frequency of blackouts in the U.S. The frequency of incidents involving the loss of external power supply at NPPs in the U.S. (based on NUREG-1032 during the period 1968–1985)

Cause of incidents involving the loss of external power supply	Number	Frequency of incidents involving the loss of external power supply (/site/year)	Median of duration (/h)
In-plant equipment failure and human error (including on-site lightning strikes)	46	0.087	0.3
Transmission grids	12	0.018	0.6
Bad weather	6	0.009	3.5
Total	64	0.114	0.6

The number and duration of incidents involving the loss of external power supply at NPPs in the U.S. (according to NSAC-144 and -147 during the period 1975–1989)

Duration of incidents involving the loss of external power supply	Shorter than 30 min	30 min or longer	1 h or longer	2 h or longer	4 h or longer	8 h or longer
Number of incidents involving the loss of external power supply	49	28	21	13	7	3
Number of incidents involving the loss of external power supply due to bad weather			13	7	6	3

In around 1975, during which the former Regulatory Guide was under deliberation, a two-wire circuit was adopted for the electrical grid system in Japan, which significantly reduced the frequency and duration of blackouts. Although it may be hard to believe, single wire circuit are still in use in the U.S. even now, resulting in frequent blackouts. I remember that there was no blackout track record data concerning two-wire circuits back then in Japan and the maximum blackout duration assumed by the U.S., which uses single wire circuits, was 8 h, which is why Japan adopted the same assumption.

The bases of the assumption that an NPP blackout would not exceed 8 h were the following three points: (1) Long-hour blackouts at NPPs in Japan, which use power transmission lines of two- (or more) wire circuit system, are inconceivable; (2) Most NPPs have multiple units, which means even if a blackout occurs in one unit, another will be able to provide electricity to it; and (3) NPPs are equipped with multiple and independent emergency power-generating equipment (diesel generators).

For example, the Fukushima Daiichi NPS had external power supply systems using two- (or more) wire circuits, which could be connected among Units 1–4 or between Units 5 and 6. Each Unit had two or three emergency diesel generators. Moreover, among these generators, each one in Units 2, 4 and 6 was an air-cooled type having diversity. It was assumed that, given these adequate countermeasures, an extended loss of power supply would never take place.

Table 5.3 Track record concerning blackouts in Japan (from commissioning of each NPP to end of March 1988)

Date	Location	Duration of blackout[a]		Availability of diesel generator		Remark
		Transmission grid	Station blackout	Availability	Load	
1979.10.19	Fukushima Daiichi NPS Unit 2	0 min	Instantaneous (15 min)	Available	Present	Fukushima trunk power line No. 2 tripped due to a typhoon, which resulted in the scram at Unit 2. Startup transformer 1S, which was shared between operating Units 1 and 2, was supplying power to Unit 1 but not Unit 2 because of the capacity shortage of the transformer. This resulted in the loss of external power supply at Unit 2.
						Later, the design of the transformer 1S was improved so that it could supply power to both units simultaneously. In PSA, this event was excluded from the calculation of the frequency of incidents involving the loss of external power supply.
1985.9.12	Shimane NPS Unit 1	1 min	Instantaneous (shorter than 2 min)	Available	Present	Lightning strike tripped the Sanin trunk power lines 1 and 2 and resulted in the loss of external supply to Unit 1.
1987.8.12	Shimane NPS Unit 1	1 min	Instantaneous (2 min 50 s)	Available	Present	As above
1980.8.27	Ikata NPP Unit 1	1 min	Instantaneous (28 min)	Available	Present	Because of lightning strike, the backup power transmission line was manually stopped. Subsequently, Ikata north trunk power lines 1 and 2 tripped, which caused the loss of external power supply to Unit 1.

Source: *Station Blackout at Nuclear Power Plants* by the Working Group for Studying Station Blackout of the Committee for Analyzing and Assessing Nuclear Facility Accidents and Failures

[a]Note: (1) Here the transmission grid means a two-wire circuit transmission grid
(2) Where automatic changeover to a startup transformer, backup transformer or emergency diesel generator (EDG) is possible, such a case is not considered station blackout (SBO) but described as "instantaneous" blackout here
The times in parentheses describe the duration of incidents involving the loss of external power supply (duration of power supply from EDG to safety systems and equipment)
(3) Duration of the external loss of power supply tends to be extended if EDG is operable, probably because the immediate restoration of external power supply is not necessary at least for the time being

As I already noted, systems capable of cooling a reactor, even without electricity, are IC and RCIC. They are designed and manufactured to work for at least 8 h, which is the assumed maximum blackout duration. Electric companies were also confident that restoring a blackout would also be possible within 8 h, which is generally the universal design time.

As explained above, policies for ensuring safety during station blackout were (1) to secure safety devices requiring no electricity for a certain timeframe and (2) to restore power supply within the timeframe. These policies are universal measures against station blackout and also have tacit global understanding. If these policies are wrong, that means blackout countermeasures taken worldwide are defective.

Incidentally, what triggered the Chernobyl accident was a test of power supply loss. It took longer to activate diesel generators in the former Soviet Union than those in western countries and their design was inappropriate against design basis accidents. In response, a test was performed to see whether the large inertia force of a turbine that has stopped generating electricity but remains rotating can be utilized as emergency power. This test triggered the Chernobyl disaster.

Back then, international cooperation to ensure the safety of nuclear power generation had not advanced as far as now. It happened during the cold war between the U.S. and Russia. Even so, any country owning nuclear power plants has been aware of the problem of lost power supply and has therefore strive to find and implement countermeasures. Everyone thought the problem had been already solved.

So why did the former Regulatory Guide not specifically stipulate a maximum timeframe for the loss of power supply and simply stated "a short period" unlike the specific timeframe stipulated by the Nuclear Regulation Authority?

In those days, in laws and government ordinances as well as guidelines and standards, phrases such as "immediately" and "as soon as practicable" were often used. These expressions are used to intentionally obfuscate a timeframe when exactly identifying it is difficult and doing so may seriously affect related matters. The expression "a short period" was used to reflect unpredictable adverse effects caused by clarifying times or periods. I learned from members who had developed the former Regulatory Guide that not clarifying a timeframe when developing a guideline was common practice back then.

Initially during the development of the former Regulatory Guide, it was highlighted that the expression "a short period" should be avoided and the timeframe clarified. Such a proposal, made to prevent ambiguity in the former Regulatory Guide, gained little support. This was because it is impossible to predict the duration of a blackout and most past track records concerning blackouts indicated durations of less than 1 s. Moreover, there was a majority opinion that assuming a permissible blackout time during an accident involving electric power companies, which are obligated to provide stable power supply in the first place, is unacceptable. This opinion is quite understandable and an allowable blackout time is more opaque than the height of a tsunami.

The reasons why the expression "short period" was adopted included the power supply situation at the time and the obligations of electric power companies.

Conversely, the Nuclear Regulation Authority (NRA) clearly requires nuclear power plants to withstand the loss of external power supply for at least 7 days and the NRA is probably demanding this after taking the Fukushima Daiichi NPS's example into account. Such a requirement may be inevitable but the former Regulatory Guide was developed after very strict deliberations and with clear bases, even while taking lawsuits into account, because it was a compulsory law and the stipulations in the former Regulatory Guide were there to prevent unexpected adverse effects. Therefore while it was possible to state a "short time" in the regulatory guide, it was impossible to require recovery from a blackout within 7 days, because its theoretical basis did not exist.

One problem when developing the former Regulatory Guide was the fact that a blackout provision clearly identifying the allowable blackout timeframe could potentially be applied mutatis mutandis to other administrative issues and become a fait accompli. If the "seven days" stipulated by the NRA became applicable mutatis mutandis and a court decision obligated 7-day blackout countermeasures for general public facilities, it would significantly cost the general public and society. Although it is foolish to spare expenses for necessary safety issues, spending on measures against excessive concerns is also a waste of government funds. Getting this balance right when developing a legally binding regulatory guide is difficult.

Incidentally, if a blackout lasting 7 days or longer took place, would the government compensate for it? Because the NRA is a government organization, its decision is the same as government policy. A blackout lasting for 7 days or longer represents a violation of the NRA's decision. According to a normal compensation claim theory, if a 7-day blackout were permitted, even for nuclear power plants, which are dangerous facilities, places where people live, which should be far safer, would naturally not have blackouts lasting for 7 days or longer, and should they occur, they could be blamed on government negligence. I cannot predict how a court would rule in such cases but this is a possible scenario.

Should a blackout lasting for 7 days actually take place in Tokyo, elevators would stop in high-rise buildings and condominiums. During a blackout, tap water would be unavailable and cooking food and flushing toilets would be impossible. People would have to climb up and down high-rise buildings and condominiums to maintain their daily lives.

It is said that many elderly people live in high-rise condominiums. Is the Tokyo Metropolitan Government well prepared for this situation? In the event of a 7-day blackout, would the safety of residents in high-rise condominiums be protected? Some people may die during such blackout. The aforementioned compensation issue would inevitably emerge.

When I consider it, all things created by modern civilization, including the city of Tokyo and nuclear power plants, are artificial structures that cannot be safely operated without electricity. Modern societies where rising populations require energy cannot exist without electricity.

The cause of the Fukushima accident was the tsunami-triggered station blackout. The key lesson learned is that station blackouts must not happen, and even if they do, their duration must be minimized.

In the former Regulatory Guide, a seemingly ambiguous expression "short time" was used in consideration of various circumstances and of compositional common practice as I already explained. While using such expression, the former Regulatory Guide strongly required the Government and nuclear industry to ensure systems and equipment would be recoverable within a short time during station blackout as well as anti-disaster measures. The Nuclear Safety Commission, which developed the former Regulatory Guide, adopted this expression, with a strong determination that "station blackouts at an NPP must be swiftly ended and not prolonged." Electric power companies were also confident of realizing it based on their past performance. However, it turned out that the requirement of the former guide could not be implemented and disasters far exceeding the strong will of the related parties occurred, resulting in the Fukushima accident.

In the new regulatory standard, "seven days" has been clearly stated. I do not think, however, that this alone will ensure safety. To utilize the lessons learned from the former Regulatory Guide, it is necessary to implement what is actually required without excessive focus on the "seven days" required by the new standard. Disaster prevention and safety measures discussed in the next chapter are keys to establishing future nuclear safety.

5.3 Layout of Equipment and B5b Issue

Most of the Fukushima Daiichi NPS's emergency diesel generators were in a basement of the turbine building. Moreover, most of the distribution panels for providing electricity to major plant equipment were on the first floor or in a basement. I heard a post-accident criticism that it was irrational to place electrical equipment which could not resist water effectively in a basement of the turbine building, which is not a waterproof structure, and it should have been placed in the watertight reactor building, which is a quite reasonable opinion. Although I do not intend to argue against such criticism, let me explain the historical background and circumstances dictating why the aforementioned equipment and components were placed in a basement of the turbine building.

Initially, the height of the Fukushima Daiichi NPS site was 35 m above sea level, but the site was intentionally excavated to about 10 m above sea level, which is its current height. The level of seawater pumps installed along the coast is lower, about 4 m above sea level.

Why were they installed so low at nearly sea level? The major reason is that the performance of machine products was not as good or reliable at the time compared to today. For example, the bearings of ordinary pumps used then were made of copper with oil grooves engraved inside. Daily maintenance and inspection meant supplying oil to the grooves and checking the pump bearing temperature by touching it with the palm to ensure no abnormality. Because operators wanted to avoid using oil-soiled pumps in the nuclear power plant to prevent radioactive contamination, pump bearings and water seals were elaborated by manufacturers. In those days, by lack of operating experiences, so pumps often failed.

In particular, most of the large rotary machines such as turbines and pumps were custom-made and each turbine blade was handmade, so the performance of machines varied. In particular, mechanical engineering frequently involved installing pumps at low places so that the pump suction would not be subject to excessive load, which was their weak point. It is wrong to criticize the former plant equipment layout from the perspective of today's high-performance rotary machines. The reason why large circulation water pumps for cooling condensers were installed near the coast is the technical background at the time.

Similarly, the original plant site was excavated to lower its level and improve the cooling of the reactor, which was the best technical measure back then.

As for the other point, why the emergency electrical equipment was installed in the basement, this was to "imitate advanced countries" in a word. It reflected scientific, technological and historical process where newly emerging academic fields such as welding, automatic control and system engineering developed, which led to the emergence of nuclear power plants, which, in return, further refined such newly emerging engineering fields. In those days, the U.S. spearheaded these engineering fields, one of which was the plant equipment layout.

The Fukushima Daiichi NPS Unit 1 was designed and manufactured by the U.S. General Electric Co. In those days, equipment layout and design by the U.S. looked smart and reasonable in the eyes of Japanese, who were still poor. The key point of the plant layout was to place the same machines and equipment in the same area side by side. It was advantageous in that pipes and tubes were placed in an orderly and compact manner, which facilitated their maintenance and kept it uniform. The basic reason why all the emergency power supply devices were installed in the turbine building was the technical background in those days. Moreover, emergency diesel generators were placed in the basements with seismic resistance in mind, because they are heavy.

The layout design of Units 2 and 3 replicated Unit 1, which was inevitable because Japanese manufacturers those days were not adequately skilled in nuclear power plant design. No-one disagreed with the belief that a tsunami could be prevented by the height of a site and the overriding opinion was that safety against natural phenomena could be ensured by a flexible design (with a safety margin) while taking into account the maximum tsunami height recorded at each site. This was the most advanced safety design method and also a universal concept at the time.

In the four decades since, Japanese nuclear power generation technology has advanced and is now regarded as the highest level in the world. There was a time when nuclear power plants were manufactured and developed by Japanese manufacturers and their safe operating records were evaluated at the highest level in the world. Such evaluation should be taken at face value. During the early 1990s in particular, those engaged in nuclear power plants in Europe and the U.S. incessantly visited Japan to see its nuclear power plants.

Although various criticisms emerged after the Fukushima accident, many countries still hope to purchase safe Japanese-manufactured nuclear power plants, despite such criticisms, which reflects its past track record. Thankfully, many countries consider Japanese technologies excellent and its products reliable.

A question was posed to conventional plant layout and design following a series of terrorist attacks on September 11, 2001. Although one of the hijacked airplanes crashed to the ground without fulfilling its purpose, it was reported that Al-Qaeda had initially planned to crash the airplane into a nuclear power plant. Taking this accident seriously, the Nuclear Regulatory Commission (NRC) focused its attention on measures against terrorism aimed at nuclear power plants, and in 2005, ordered each national nuclear power plant to reinforce and disperse emergency power-supply devices. This order is known as B5b, which is the article number in the order document.

The order B5b was notified to each country having nuclear power plants as a top secret recommendation and each country secretly carried it out. Certainly the recommendation was also notified to Japan. The Japanese government, however, did not communicate it to us engaged in the private nuclear sector and shelved the notification, meaning that Japan regrettably lost its best opportunity to prevent the Fukushima accident.

Had this recommendation been communicated to nuclear-related parties, emergency power generators would have been added and dispersed and power supply would have been available from sources unaffected by the tsunami, preventing the Fukushima accident. Although the Fukushima Daiichi NPS Units 5 and 6 are located in the same site and suffered the same tsunami disaster as other units, both were brought to cold shutdown by using only one emergency diesel generator having survived the tsunami. The operators of Fukushima Daiichi NSP had such a proud operating technique.

Concealing the B5b meant the loss of the best and last opportunity and I think this is the main reason behind the expansion of the Fukushima accident. In this regard, the Government is more liable for the accident than TEPCO. Moreover, this concealment is almost a crime rather than a mistake or dereliction of duty.

My lecture seems to have heated up, so let us take a break here. Although some may wonder whether the dispersion of emergency power supply devices recommended by the B5b may compromise maintenance convenience and precision, such concerns are misplaced. During the past 30 years, mechanical engineering has progressed significantly. Although pumps used to require annual overhaul day, they now have highly reliable structures, such that any overhaul may itself compromise their precision. Repairing a pump while greasing is now unheard of.

5.4 Reliability and Diversity of Equipment

In the 1970s when the Fukushima Daiichi NPS started operation, large passenger aircraft B747s known as "jumbos" went into service, which were capable of carrying more than 400 passengers and had four jet engines. It was said that even one engine would be enough for them to fly.

Thirty years later, their role ended due to the oil price hike and changing passenger numbers after local airport development, and in the 2000s, B777 aircraft began to fly. A B747's fuselage length was 70.6–76.4 m whereas that of a B777 is slightly shorter at 63.7–73.9 m with the number of engines reduced from B747's 4–2.

Aircraft are machines carrying people's lives and the reliability of their engines equates to their overall reliability. The fact that the number of aircraft engines was halved from four to two means their reliability has more than doubled and airlines, which are responsible for passengers' lives, have recognized it. Although it may have been generally gone unnoticed, technological quality, namely product reliability, has dramatically improved thanks to the development of the mechanical industry over the past 30 years.

Reviewing the construction period of the Fukushima Daiichi NPS 30 years ago, the only reliable machines that could be used as emergency power generators and which were required to operate for an extended period were water-cooled diesel generators. Accordingly, installing water-cooled diesel generators redundantly was inevitable.

This must have been a big problem in preventing terrorism for the U.S., which owned many old-design nuclear power plants and why the U.S. ordered the dispersion of emergency power generators through the B5b. As with tsunamis, terrorism can trigger dysfunction of multiple sets of equipment and systems with an armed attack, which is technically termed a common cause failure (CCF).

An earthquake or fire is also highly likely to trigger a CCF. Concentrating the same machines in the same place increases the potential for their being simultaneously destroyed by a common cause such as an earthquake, fire and terrorism. Even safety equipment and systems laid out in double or triple are highly prone to being simultaneously destroyed by a CCF, which is exactly what the tsunami that caused the Fukushima accident was.

Anti-CCF measures include the dispersed layout highlighted by the B5b and also diversity. Diversity does not mean installing multiple equivalent machines but rather installing different machines with the same function. It is like a car equipped with a brake and an emergency brake and it is apparent that diversity helps prevent CCFs.

Diversity is also effective against terrorism. This is because terrorists, unfamiliar with plant sites, would consider equipment and components of different shapes and sizes to differ, which is likely to reduce the potential for their complete destruction. A dispersed layout (dispersal arrangement) suggested by the B5b is a safety measure resembling diversity.

In fact, the former Regulatory Guide also required diversity.

Safety equipment cannot be deemed such if there is no other equivalent safety equipment installed independent of the same. This is called redundancy. Conventional wisdom of redundancy assumes that if two equivalent and independent systems exist, even if one is damaged, the other will survive to achieve its objective. However, the redundancy requirement concerning emergency power supply, as stipulated in the former Regulatory Guide, had a proviso to require redundancy "as much as possible." This suggests at least the need for diversity in safety design has long been recognized.

When the Fukushima Daiichi NPS was constructed, it was impossible to adopt diversity, even though its necessity was perceived. Back then, the only reliable emergency power supply devices were water-cooled diesel generators. Accordingly, the only way to comply with the former Regulatory Guide was to provide water-cooled diesel generators redundantly.

Nowadays the situation has changed and it is easy for safety designs to incorporate diversity. For example, air-cooled diesel generators and gas turbine generators are available and the time required from activation to full power has shortened. Many products capable of meeting the design requirements of nuclear power plants exist, where a whole range of equipment and devices is used. Today, emergency generators need not necessarily be water-cooled and a wide variety of options are possible.

Regrettably, however, as long as 40 years elapsed since the Fukushima Daiichi NPS was constructed without such technological progress being incorporated. TEPCO is liable for not having adequately incorporated the diversity required in the proviso of the former Regulatory Guide. This shortfall also went unnoticed due to negligence by the regulatory authority. Both are liable.

5.5 Natural Disasters and Safety Design

The meaning of "nuclear safety" has been gradually changing.

Originally, NPP safety designs were intended to prevent initiating events such as occurred by operator mistakes e.g. during maintenance and inspection, or a mechanical failure, and provide mitigation measures (safety systems and equipment) to protect the reactor appropriately by design.

In safety designs, natural phenomena such as earthquake, tsunami and typhoon were handled as a failure triggering initiating event(s). It was thought such a failure could be eliminated by constructing buildings and seawall based on past worst data and designs with an adequate safety margin. Unexpectedly, the Fukushima accident made us realize such views were wrong. As well as being an initiating event, the tsunami also has the aspect of a threat induced common cause failure that completely destroys emergency power supplies.

An event destroying all emergency power supplies is the same as terrorism. What tsunami and terrorism have in common are two points, first a destructive force beyond design (beyond anticipation) and second the capability to disable any safety system and equipment provided with redundancy and independence (common cause failure).

Conversely, current NPP safety designs are intended to secure entire plant safety, even in the event of an accident where plant systems and equipment are destroyed (initiating events) and the most effective mitigation measures (safety equipment) cannot be used (single failure). In Japanese language, safety design for building nuclear power plants and safety measures against common causes such as natural phenomena and terrorism are both expressed with the word "safety" but they are quite different.

Their difference is easy to understand in comparison to traffic safety for example. Plant safety design is equivalent to the design for manufacturing a high-quality car. Conversely, safety against terrorism and natural phenomena is equivalent to overall social traffic safety, including roads, road signs, and appropriately arranged gas stations, based on which cars can run safely. Thus the concept of safety has changed these days from plant safety to traffic safety.

Comparing the safety required for NPPs and traffic safety, NPP safety design takes into account factors resembling those for safe driving such as measures against uneven gravel roads, appropriate gas station intervals and rear-view mirrors. The single failure rule guarantees NPP safety design more than traffic safety rules guarantee driving safety. However, as cars are not designed to run on desert sand and require oil supply at gas stations at appropriate intervals, NPPs are also designed assuming sound social conditions basically, which is more or less inevitable.

The Fukushima accident can be compared to a traffic accident, where roads were destroyed by an earthquake (NPP blackout), gas stations were rendered unusable due to the tsunami (station blackout) and the engine stopped (stop of core cooling), and although helpless drivers started to walk toward their destinations (starting up the decay-heat-driven cooling device), but ending up by the lack of water and food (stop of the cooling device), and fell down (core melt and explosion).

Conventionally, safety designs assumed a scenario where "drivers could walk toward their destinations unaided, even in the worst case, albeit for a short time (8 h)," and were assuming "a social structure where the necessary water and food would be available during the short time." It was not expected that a tsunami would destroy all assumed social conditions. There was a precondition that "it would be possible to exploit social conditions to a certain extent and buy water and food on the way to the destination (electricity restored within 8 h of the blackout)."

The cause of this mistake was the precondition that safety designs could cope with natural phenomena. In the Fukushima accident, natural conditions caused a common cause failure destroying the safety systems and equipment entirely, which is a significant mistake to be corrected and a lesson to be learned.

A substantial difference exists between NPP safety design and safety measures against natural phenomena and terrorism, in that the former must cope with initiating events and the latter on common cause failure induced by threat. Safety measures must be developed after modifying the current preconditions first.

Although it may sound complicated, the safety of nuclear power generation means comprehensive safety, taking into account terrorism, natural phenomena and the safety measures to treat differ from mere safety design. The safety of nuclear power generation is comparable to e.g. traffic safety comprising roads, road signs, gas stations, police and the JAF (Japan Automobile Federation), etc. Safety design is more limited like the design of cars and differs from the comprehensive safety of nuclear power generation.

With this in mind, what is necessary to sublime safety design to the level of nuclear power generation safety? The answer is disaster prevention and safety measures, which I will discuss in the next chapter.

Chapter 6
Reconstruction of Safety

6.1 Conventional Nuclear Safety

I have discussed the core melts and explosions, releases of radioactive materials, tsunami and station blackout at the Fukushima Daiichi NPS. In this chapter, I am going to discuss the history and transition of nuclear safety. Based on reflections on the past nuclear safety, I am going to propose a direction of safety measures to be taken in future and in so doing, point out mistakes made by the Nuclear Regulatory Authority (NRA) and many mass media. Although it may sound argumentative, I ask your understanding. General readers can also skim through this chapter and grasp only its outline.

This is because fission reaction is so fast and generates so much energy that it was considered impossible to control nuclear reactors through a past approach by human eyes and hands at the beginning of the development.

The technological level at that time can be understood by considering the circumstances at WWII. Soldiers aimed at targets using telescopes and fired cannons with commands. Today, long-range missiles automatically reach their targets with an accuracy unimaginable in old times. Nuclear power generation emerged at such an early stage of technology that accurate and ceaseless automatic control, in place of human operation, was deemed indispensable. Leading countries began to engage in fierce competition in the research and development of nuclear power generation.

The first manufactured commercial NPP was the Calder Hall NPP of England. Next, the U.S. built the PWR Shippingport Atomic Power Station. Subsequent light water reactor development was promoted by the cooperation among the U.S., Germany and Japan. The current safety designs of PWR and BWR would never have been achieved without the cooperation of these three nations, resulting from the energy crisis due to the oil shock in the early 1970s.

The reason why light water reactors have become the most popular global option is their proved safety designs compared to other types of reactors. The IAEA's

© Springer Japan 2015
M. Ishikawa, *A Study of the Fukushima Daiichi Nuclear Accident Process*,
DOI 10.1007/978-4-431-55543-8_6

Nuclear Safety Standards (NUSS) were based on the safety design guidelines developed by the U.S., Germany and Japan with devotion and enthusiasm.

Nuclear plant control and safety ensured by machines, which was the fundamental requirement in the early period of nuclear energy development, has reached perfection. You may well ask at this point "Then why did the Fukushima accident happen?" I am now going to answer that question.

Although present standards have been reconfigured, the IAEA's international safety standards NUSS included not only safety design guidelines but also five other guidelines related to governing organizations, siting, operation and quality assurance. Incidentally, the Chernobyl accident happened only one year after completion of the former safety design guideline (NUS) finally developed by the IAEA's safety advisory panel (SAG) after decade-long deliberation. The NUSS was developed by revising the NUS based on reflection on the Chernobyl accident and is commonly known as the revised IAEA safety guideline 1992.

This revised guideline changed the safety system from the one with total dependence on machines (safety design) to the partial allocation of responsibility to operation management based on lessons learned from the TMI and Chernobyl accidents caused by a series of human mistakes. Roughly speaking, two significant modifications have been made. One is to provide a greater safety margin for the design pressure of containment vessels and the other is to prepare evacuation routes of local residents prior to commencement of operations.

To be more specific, these two modifications were based on the acknowledgement of severe accidents, the existence of which had not previously been taken into account in safety design, and requested addition of necessary safety measures. This underlines the need for emergency measures against severe accidents and the potential to mitigate severe accident conditions via such measures.

The safety margin added to containment vessel pressure design also means operators will have more time for emergency operations.

This means that the safety concept has shifted from the initial development stage, where reactor control depended on machines, to the one where the final responsibility for safety lies with humans. Accumulated experiences of nuclear power generation have reaffirmed the need for manual and visual reactor control, particularly during an emergency. Operators' actions to mitigate accident situations are called accident management, in which the safety roles of operators are predetermined. Ultimate responsibility for accidents has shifted from machines and equipment to operators, which marks a significant shift in the safety concept.

Ensuring evacuation routes well before the first criticality means that the potential for severe accidents exceeding safety design (safety functions of machines and equipment) has been recognized, suggesting the potential for disasters requiring the evacuation of residents.

Since the Nuclear Safety Standards (NUSS) have acknowledged the potential for severe accidents, countries owning nuclear power plants have started taking relevant measures. The vent, a key subject in the Fukushima Daiichi accident, is one such example. These vents, which had not been incorporated in a BWR safety design previously, have since been adopted as a measure against severe accidents.

Looking at the papers presented at the international conference in these days, there were a variety of proposals against severe accidents, and most of them were related to the safety margin of a containment vessel. However, proposals for specific severe accident countermeasures were scarce and each country took measures while taking into account their own circumstances.

This is because there was no example of LWR-related severe accident except the TMI accident, a situation which made it impossible to take specific severe accident countermeasures. Specific severe accident examples which could prompt measures were lacking. In this sense, the Fukushima accident is a negative lesson that highlighted specific problems in terms of nuclear emergency readiness to the world.

Experts who gathered at the IAEA had no specific severe accident countermeasures except securing evacuation routes in advance. While overestimating the potential for severe accidents will waste national funds, underestimating the same may result in a tragedy. TMI-related findings and knowledge were insufficient to develop severe accident countermeasures. More findings were in demand, which was the situation in around 1992 and it is most regrettable and ironic that they were provided by the Fukushima accident. Because I was the representative of Japan in the IAEA Nuclear Safety Standards Advisory Group (NUSSAG) for about 10 years, I remember the atmosphere in those days very well.

Changed mindsets in the 1990s, namely a shift in focus from safety design to operational safety and from measures against design basis accidents to those against severe accidents, were right and timely. However, emergency measures against severe accidents were vague and inconclusive for each country, which hampered their efforts to develop them and saw repeated trial and error, reflecting the global trend at the time.

One example is the comprehensive nuclear disaster readiness drill, previously a regular event in Japan and the Prime Minister on TV screens wearing a drill uniform and taking command in drills simulating NPP accidents was a familiar scene which emerged during such period. Despite the TV appearance of the Prime Minister, the development of specific severe accident countermeasures, which were the essential point, did not progress at all, except for installation of vents by electric power companies. In those days, disaster prevention awareness-building campaigns were the best that could be done.

A safety myth frequently spouted by mass media describes this situation perfectly. In TV programs, experts, antinuclear activists and diet lawmakers insisted on ridiculous story with a smug look, that nuclear accidents would kill millions of people, while no-one discussed what measures should be taken. Considerable time elapsed without developing any specific severe accident countermeasures, resulting in the tragedy of March 11.

6.2 Safety Design and Natural Disasters

What did the Fukushima accident suggest to us? It left a specific lesson that a natural disaster should not simply be regarded as an input (requirement) for safety design but as a real threat that may trigger a severe accident. It taught us that the destructive force of nature is one capable of exceeding safety design. The same is true of overlapping operational mistakes and terrorism.

In conventional safety designs, natural phenomena were considered design conditions based on which robust structures should be constructed. Accordingly, past history was researched, involving the identification of maximum records and frequency and possible worst conditions (scenarios) determined, based on which (reference values) buildings, machines, and equipment have been designed and manufactured. This is a universal concept. Although, there are innumerable natural phenomena such as earthquakes, tsunamis, typhoons, floods, tornados, volcanic eruptions, heavy snowfall, blistering heat and frigid climate, each of which has been handled as a design condition to be conquered with design considerations.

Exceptionally, special design methods were established against earthquakes among these natural phenomena, namely seismic designs. The worst conditions were input into the seismic designs of research reactors in the earliest days of nuclear power generation. This was a relatively easygoing age where designs anticipating earthquakes of a magnitude three times that of the Great Kanto Earthquake were deemed adequate to ensure safety. The 10-m height of the Fukushima Daiichi NPS site was also regarded as safe under a similar mindset, reflecting the knowledge of natural phenomena at the time.

Since then, however, earthquake-prone Japan has learned from U.S.-engineered seismic design methods, such as seismic wave detection and analysis, and seismic ground motion analysis based on the same. Consequently, plant buildings became strong to withstand earthquakes, a change seen around in 1975. Also in the Great East Japan Earthquake, all 15 NPP facilities located around the earthquake source area, including the units of Fukushima Daiichi NPS, were resistant to beyond design basis earthquakes with a magnitude of nine.

The National Diet of Japan Fukushima Nuclear Accident Independent Investigation Commission reported a pipe rupture due to seismic motions may have occurred at Unit 1 of the Fukushima Daiichi NPS, although there was no evidence to support it. With regard to the other 14 units affected by the Great East Japan Earthquake, despite severe tsunami damage, no safety-related structures, systems or components were damaged. I really wonder why Unit 1 alone is suspected of pipe rupture due to the earthquake. During the 40 min until the tsunami hit, the Unit 1 reactor stopped normally and operated toward a cold shutdown smoothly. All plant data were also normal with no data indicating earthquake damage. The investigative committee's report, which was written while disregarding this fact, seems to have been intentionally distorted.

To return to the original subject, these seismic design methods were established after seismic wave analysis started and achieved by engineers targeting manufacturing, by participating in the field of earthquake, namely geophysics, and fully exploiting engineered methods. Earthquake waves of all sorts were collected and analyzed, whereupon all-out seismic ground motion analyses were performed and quake-resistant designs of building structures, machines and equipment were established. The seismic isolated building that played a key role during the Fukushima accident is one such achievement.

Seismic design can represent a benchmark for safety measures to be taken against all natural phenomena. The former Regulatory Guide for Reviewing Safety Designs of Light Water Reactor Facilities (hereinafter referred to as the safety design guides) stipulates that the worst estimates concerning all natural phenomena, excluding earthquakes, must be used as benchmarks for designs, because no other valid method has yet emerged and this is a global problem to be tackled.

Similarly as the hazardous factors of an earthquake have been successfully identified as seismic motions, there is a need to clarify and identify the hazardous factors of all natural phenomena, collect broad data, and develop means to resist such phenomena. As it was successful in seismic designs, engineered investigations into the threats of all natural phenomena should be made. Despite the huge challenges such work imposes, it is achievable within 20 years through international cooperation centering on the IAEA. Such thorough investigations would surely result in NPPs capable of withstanding natural phenomena and the investigative results could also be broadly used in other industries.

These are methods of building NPPs to withstand natural phenomena, in other words measures against natural phenomena to be considered in safety designs. Stronger NPPs would prevent the repetition of the Fukushima accident tragedy, where all functions of safety facilities were lost at once due to the tsunami. Such a reactor safety design itself is passable at least.

However, what must be taken into account is the lack of any guarantee that a natural phenomenon exceeding such safety design will never take place. Although some may rule out such concerns, events exceeding such a robust safety design should be anticipated. This is a lesson that should be learned from the Fukushima accident. An additional layer of mitigation safety system against the threats of natural phenomena should be provided based on reflection on the Fukushima accident.

A case in point is the long-hour loss of external power supply due to a tornado at the three units of the Browns Ferry Nuclear Power Plant on April 27 in 2011, immediately after the Fukushima accident. This blackout lasted more than 8 h, which was the maximum permitted outage time constituting the basis of the former safety design guides. It lasted so long that President Obama worried and rushed to the plant to encourage operators. Consequently, emergency power generators whose performance had been reinforced because of the B5b operated for extended hours and brought the reactors to a cold shutdown state.

It is said that in the U.S., blackouts due to hurricanes occurred sometimes after this blackout. These days, violent natural phenomena seem rampant.

As I stated previously, natural phenomena should not be regarded as merely inputs for plant safety design but also as threats capable of destroying all safety facilities, based on which measures should be taken. Such measures are mitigation safety systems against disasters (MISSAD).

6.3   Mitigation Safety Systems Against Disasters (MISSAD)

No measures can completely rule out natural threats attacking an NPP. This is apparent from the fact that the Great East Japan Earthquake and the subsequent tsunami caused about 20,000 fatalities and disappearance. Before the Great East Japan Earthquake, the Seismological Society of Japan had initially concluded that no earthquake threat existed in Fukushima prefecture's Hamadori area, where the Fukushima Daiichi NPS is located, which shows the limitation of science and technology.

Accordingly, we have no choice but to thoroughly investigate the damage and to strive in mitigating the effect of future disasters can be mitigated.

Although the Fukushima accident caused more than 100,000 evacuees as well as core melts and hydrogen explosions, no-one died due to the reactor accident and radiation disaster. This nuclear disaster killed no-one. The staff and related parties of the Fukushima Daiichi NPS accomplished their duties well. In this book, I call such accident response activities the "mitigation safety system against disaster (MISSAD)."

Although the MISSAD and safety design complement each other, they are completely different concept. When an accident exceeding the safety design scope happens, how can the hazard be minimized? The first action is to respond to the "severe accident." When the hazards exceed the range of severe accidents responses and activities to be conducted are called MISSAD. I am going to discuss the specifics of these systems.

After the Fukushima accident, utilities nationwide added emergency power generators and power source cars as well as constructing seawalls and filtered vents.

Of these, only the power source car falls within the range of the MISSAD. Adding emergency power generators is enhancing the safety design. A filtered vent may also be included in severe accident countermeasures if deemed as a part of the containment vessel. It can also be categorized in the MISSAD. A seawall is rather harmful and wasteful without significant merits and does not deserve the name of safety.

Does this explanation help you understand some of the differences between MISSAD and safety design?

MISSAD are intended to reduce damage caused by disasters and utilize wide-ranging means and methods in countless combinations to achieve this goal. Such measures cannot be accomplished by machinery and equipment alone, which operate just as programmed. They differ from machinery and equipment in nature. MISSAD are initiated under human judgment and actions and are totally opposite to automatically operating safety facilities.

Preparations for MISSAD also vary worldwide, unlike NPP safety design guidelines. Just as NPPs in Kyushu do not require snow and ice control measures in Hokkaido, it depends on the site environment of each NPP. In this sense, nothing is uniform, or rather everything is special.

So, what is required for disaster readiness? While this is a difficult question given that the targets are such as gigantic natural phenomena, a minimum of water, food, electricity, gasoline, designated emergency meeting places and electric and mechanical machines to facilitate emergency work are necessary. Instruments and machines used for MISSAD are also valuable auxiliary tools that effectively support human activities. One hammer and one nail can be useful for MISSAD while a seawall costing 100 billion yen can be a cumbersome obstacle. In fact, in the Fukushima accident, the most urgent needs were for electricity and cooling water. The seismic isolated building was constructed based on lessons learned from the Chuetsu Offshore Earthquake in Niigata prefecture in 2007 and proved quite useful in the Great East Japan Earthquake. The NPP staff lodged in this building to respond to the earthquake originated accidents. Staff of the Onagawa Nuclear Power Station, for example, used an earth-moving machine, which they luckily found on site, to make a makeshift road and escaped from the station. This is because the road leading to the power station had been blocked due to the earthquake.

As these examples suggest, the required equipment and materials for the MISSAD differ depending on the time and place. Although instruments and tools that may be urgently required should be stocked in a plant, a complete set for the MISSAD need not be provided in the plant. Goods not urgently required can be procured from areas not affected by the disaster. However, it is essential to establish a plan for mutual procurement in advance. Almost no other countries have developed as many transportation means as Japan, including rail, marine, air and road transportation. In addition, a heliport in a plant would be an effective means of mitigating hazards, rather than facilitating visits by the Prime Minister.

Apart from equipment and materials that should be stocked in plants, there are items that should be prepared by the Government: for example, a high-speed large power source vessel. If only a power source vessel had been dispatched to the Fukushima Daiichi NPS, the situation there would have been very different. The aftereffects of tsunamis subside within half a day, meaning such a vessel could have been brought alongside the coast in front of the NPS. Even if electricity it transmitted had come one day later, it would have prevented the core melts at Units 2 and 3. A power source vessel could have carried large pumps and emergency civil engineering machinery and equipment.

As a person engaged in the nuclear energy field, I recommend a nuclear-powered ship as the most appropriate large power source vessel. The heavy reactor loaded on it lowers its center of gravity and stabilizes it. A robust large nuclear ship can carry thousands of residents if necessary and could also be used as an emergency control tower where workers could sleep.

When a power source vessel is not used during normal times, it could be used as a training vessel for personnel of, e.g. the Self Defense Force or the Japan Coast Guard. If such a vessel were constructed in advance, it would contribute significantly to disaster relief. I remember that the Antarctic research vessel Soya was saved by the former Soviet's ice-breaking ship Ob.

During the Fukushima disaster, former Prime Minister Kan declined a relief offer by the U.S., which was a very frivolous act. He may have been arrogant as the Prime Minister of a technological superpower but knew too little about nuclear energy. Although I do not know what the U.S. had planned back then, had the former Prime Minister let the U.S. know about the desperate attempt under station blackout, it might have at least considered supplying electrical power from a nuclear submarine. In the first place, why did no-one in the Government have the idea of providing emergency support activities by sending ships of the Self-Defense Force and the Japan Coast Guard to the Fukushima Daiichi NPS to supply electrical power for it. This shows that the government leaders completely lacked a vision of MISSAD.

The issue of MISSAD is the one Japan must seriously tackle as a country prone to natural disasters. Here is some advice. Since these systems are developed and operated by humans, it is important to train people. Since the disaster response involves risky and dangerous work no-one likes, the key is to develop human resources with dedication and courage. This is the root of safety culture. I will leave the studies of specific measures to young people who bear this country's future.

6.4   Safety of Nuclear Power Plant

Now that I have explained MISSAD, I must return to nuclear safety, which is a stage preceding the MISSAD. The reason why I have delayed discussing nuclear safety is the complicated nature of the issues involved.

More than 40 years have elapsed since full-scale nuclear power generation started and the only nuclear disasters that have taken place during this period were the TMI and Chernobyl accidents. Except from the Chernobyl NPP, where a reactor type and safety concept are quite different from LWR, the TMI accident had been the only LWR accident and it caused no radiation exposure and did not kill anyone. It was just when the perception of safe nuclear power generation was about to spread globally reflecting such historical facts, the Fukushima accident occurred.

Then how has LWR safety been ensured? Safety designs are the core of the safety. In other words, a system ensuring safety depending on machines and equipment has been established. Now I am going to detail the safety design, the core of the safety by the equipment and devices.

Incidentally, the three greatest progresses in the twentieth century were nuclear power generation, space development and computers. Among these, space development depends on quality control technology to ensure the quality of machines and equipment. It would be impossible for a rocket to go into outer space were it loaded with heavy safety equipment as in the case of NPPs. The safety of a rocket is ensured by non-redundant perfect design. Conversely, nuclear safety is ensured with many necessary safety facilities. The safety philosophies of space development and nuclear power generation are opposing and contrasting.

In the prologue of this chapter, I wrote that "accurate and ceaseless automatic control was deemed indispensable for nuclear power generation." This is because nuclear fission reactions are far more rapid and generate far more energy than chemical reactions, with which human beings have been familiar since the old days. Nothing other than the accurate and ceaseless operations of machines and equipment was deemed capable of precisely controlling nuclear reactions.

Nuclear development started when the WWII aftereffects subsided, and the remarkable engineering development and progress successively took place. Nuclear power generation progressed through engineering advances and rapidly approached maturity. At the same time as the construction of nuclear power plants in public areas started, ensuring their safety became an issue. Provided there was no choice but to depend on machines to control nuclear fission reactions generating enormous energy in a short time, safety was inevitably dependent on machines. The concept of safety design practically started in around 1960 and was triggered by the SL-1 accident that occurred in the U.S.

Nuclear safety means constructing NPPs free from radioactive disasters. Specifically, it means designing an NPP so that barriers that confine radioactive materials, including fuel rods, can be protected without fail.

In the case of LWR, three barriers are provided to confine radioactive materials, namely cladding tubes for packing nuclear fuel, which causes fission reactions, a primary coolant system, where water for cooling fuel rods circulates, and a containment vessel, which envelopes nuclear power generation systems and equipment. These are technically termed fuel cladding, reactor coolant pressure boundary and containment boundary respectively.

Safety design is intended to construct an NPP free from radioactive disaster by protecting radioactive fuel rods and radioactivity-confining barriers without fail. The containment boundary is a robust pressure-proof structure intended to function as the final barrier for confining radioactive substances in case the other two barriers have been ruptured.

In the earliest days of development, no technology existed to build containment vessels as large as those of today. Accordingly, the Shippingport Atomic Power Station, the first constructed PWR plant, had as many as three containment vessels. Among these containment vessels, one housed a reactor and the others housed steam generators. A turbine power generator was also installed on the top of a building that housed a condenser with coarse roof. If it remained today, it would be designated as a World Heritage monument. Regrettably, it was dismantled and removed as the first decommissioned plant and the site is now a vacant lot.

In those days, fixed NPP designs did not exist.

The safety design starts by examining the potential malfunction or breakdown of facilities used in a power-generating system, ranging from reactors to turbine power generators. The first malfunction or breakdown is called the initiating event, which also encompasses incorrect human operation or faults in maintenance or repair. The initiating event makes a power-generating system having something deviated from normal operating condition. For example, damage to a coolant pump results in abnormal conditions such as an increase in fuel rod temperature and reactor pressure. Systems and equipment that prevent or mitigate such abnormal conditions are called safety facilities.

In the safety design, the facilities drawn on design blueprints are destroyed one by one as potential independent initiating events, their effects on the aforementioned three radiation barriers are examined and the most appropriate safety facility is designed as a countermeasure. Naturally, since there are many initiating events that result in similar abnormal conditions, the required safety facilities tend to resemble each other, although they may differ in size. Of the candidate facilities for the same application, the facility having the largest capacity is adopted as the safety facility for this application, in line with the theory "the bigger, the better." If all facilities in the power station are examined like this, the resultant group of safety facilities would be theoretically capable of coping with any initiating event.

Such safety design check is the role of safety analysis. Calculations and analyses are performed for each initiating event to check whether safety facilities function properly so that fuel rods and pressure boundary will remain undamaged. Furthermore, single failure assumption is added to this operation to make it stricter. Single failure assumption is a quite "demanding" rule because it intentionally excludes safety facility which is deemed most effective in recovering or mitigating abnormal conditions caused by an initiating event. To overcome such a severe single failure assumption, there is no choice but to install plant safety facilities redundantly.

Safety design guidelines require the redundancy, independency and testability of safety systems to be equipped in an NPP based on such single failure assumption. This is why electronic power companies emphasize they have two- and three-layered safety measures.

To check whether the analytical results are correct and the integrity of fuel rods and safety barriers will be maintained, verification through experiments is indispensable. Because the target of such experiments is a reactor, and the reactor is in an accident condition, they are not easy. The U.S. performed such experiments by building actual test reactors in vast deserts, in which Japan and Germany cooperated. This marked the beginning of international cooperation for nuclear safety research, which started in the early 1970s.

Following the reactor experiments in the U.S., actual accident situations in a reactor was grasped comprehensibly. The PCM experiment mentioned in Part I, Chap. 1, Sect. 1.3 is one such achievement. Japan and Germany classified accident behavior to implement supporting experiments and analyses for each type of behavior, the results of which were integrated to develop and verify safety analysis codes, which was completed before 1980.

The results of these experiments and analyses effectively correlated what were generally complicated reactor accident phenomena. All related parties believed that provided an NPP were built in accordance with safety design, its safety would be ensured. However, the TMI accident occurred soon after in March 1979. Although its cause was pump failure, overlapping maintenance errors with operators' judgment errors resulted in a core melt. As I noted previously, this accident triggered a quest for safe operation.

Safety design is very effective, working almost perfectly against simple failure and system malfunctions. However, safety designs fail in the event of successive incorrect human operations, causing a complicated situation exceeding the single failure rule as in the case of the TMI accident. Under such circumstances, safety operation by operators who can grasp the situation effectively become necessary, as so-called accident management (AM).

Safety design is intended to provide cutting-edge systems and equipment for protecting nuclear safety but has not emerged as the almighty safety means which was expected during the early nuclear development period. Since then, safety design has lost its position as the primary means of ensuring nuclear safety and the onus has shifted to NPP safety management supervisors.

Moreover, the need to modify the method of evaluating natural phenomena in safety design should have been recognized at that point. During the revision of the NUSS safety standard in 1992, all the attention of related parties was directed toward safe plant operation while they regarded natural phenomena as simply inputs for building robust structures and did not consider them threats capable of totally destroying safety facilities. The Fukushima accident directly hit this weak point and made us realize that natural phenomena are threats capable of exceeding the capacity of safety design.

However, safety design is not at all utterly helpless in the face of natural phenomena. In the Fukushima Daiichi NPS Units 5 and 6, operators managed to utilize the only emergency power generator that survived the tsunami to achieve cold shutdown. The emergency power generator was one of the safety facilities manufactured according to safety design.

Stronger evidence showing the importance of safety design is the difference in the time of core melts among Units 1–3. The decay heat operated reactor core isolation system (RCIC) in Unit 2 and 3 postponed core melts there, as long as they functioned. This example shows that safety design remains a foundation for protecting nuclear safety.

During the early period of nuclear power generation, safety design and nuclear safety were synonymous. However, such a total reliance on safety design triggered the Fukushima accident. This fact underlines the strong need to evolve from safety design to nuclear safety. This can be compared to the automobile society, where safety is ensured not only by car design but also by broad means and systems ensuring traffic safety. While safety design is certainly important to ensure NPP safety, solving the issues surrounding NPPs, which are comparable to broad traffic safety issues, is crucial.

Nuclear safety designs, like car designs, are universal as represented by the IAEA safety standard NUSS. NPPs are sometimes designed with site-specific topography and social circumstances in mind, but reactors and power-generating facilities and equipment, which are most crucial in terms of safety, are designed universally. NPP designs are generally the same. What make NPPs differ from one another, if at all, are whether they are new or not, built well or not, and have good quality or not. Nowadays, most reactors are LWRs and therefore NPP safety designs are generally the same worldwide. As any car in the world has its main and emergency brakes, the safety designs of LWRs operating in the world are generally the same.

Although the Nuclear Regulation Authority (NRA) has declared it will enact and enforce new and globally unrivaled safety guidelines (regulatory standards), they seem overly worked up. Their current regulation with excessive focus on active faults is not a nuclear safety regulation. Regulations by the now defunct Nuclear and Industrial Safety Agency (NISA) stuck to the details of quality assurance and were nicknamed "omission and error" regulations. Neither the NRA nor the NISA seems to have rationally performed the safety design concept I have discussed.

The new regulatory standards are studded with dubious definitions and classifications. The NRA should note the fact that as the designs of cars used worldwide are generally the same, the same applies to NPP safety designs.

Accordingly, to ensure nuclear safety, the safety of NPP surroundings, in other words, MISSAD, which have not been adequately considered in safety design, must be ensured rather than the mere modification of reactors (safety design). MISSAD are more comprehensive than safety design and can be compared to roads for cars, traffic signs, service stations located at appropriate intervals, and the JAF dealing with accident and disabled cars as well as other traffic safety systems.

6.5   Antiterrorism Measures

6.5.1   What Is Nuclear Terrorism?

Although not a specialist in the field of antiterrorism measures, I was a member of an IAEA committee, the Advisory Group on Nuclear Security (AdSec) for 2 years as an expert in nuclear reactor safety. For nuclear safety, measures against terrorism will become important to ensure nuclear disaster prevention and safety, in addition to those against natural phenomena. While admitting my inexperience in the field of antiterrorism measures, I am going to discuss them.

First, nuclear terrorism can be described as "4×4," which is a method of easily memorizing its types and covers all types of terrorism. The first "4" refers to the following acts:

(1) the theft of nuclear weapons
(2) the theft of nuclear materials
(3) the theft of radioactive waste, and
(4) attacks on NPP.

For terrorists, obtaining a nuclear weapon would be the quickest way to implement terrorism, which corresponds to item (1). However, stealing a nuclear weapon is very difficult, even for a terrorist organization, due to strict guards against terrorism. Instead, taking over a country possessing nuclear weapons might be easier, which is why the U.S. government is nervous about politically unstable countries possessing nuclear weapons.

Item (2) relates to physical protection (nuclear material protection). Physical protection is the IAEA's main activity and has been performed continuously since its establishment. Following the collapse of the former Soviet Union, the IAEA very nervously monitored plutonium and other radioactive materials stockpiled there to prevent their outflow overseas.

Item (3) seems easier than (1) and (2), but its effect would be smaller.

Radioactive waste can be classified into high- and low-level waste. High-level waste is produced by vitrifying radioactive effluent generated by reprocessing spent nuclear fuel and extracting uranium and plutonium from the reprocessed spent nuclear fuel. Such waste is also called vitrified waste, whose very high radiation dose is such that directly watching it nearby causes radiation exposure equivalent to the JCO criticality accident, which instantly knocked out or killed JCO staff members. Vitrified waste is transported in shielded containers under strict guard, which even mysterious thief Arsene Lupin would be unable to steal. Even if it were stolen, opening its container would be impossible without a large hot cell and opening its container without protective gear would result in instant death.

Conversely, low-level waste is generated in bulk but its radiation level is not so high and a bomb containing such waste is called a dirty bomb. Some people worry dirty bombs may be used by terrorists, but their radiation level is low and thus their effect on human bodies is essentially low. Although it may be extreme to say that their effect can be discounted, devices capable of measuring even tiny amount of radioactivity are available these days, so even if they were used, contaminated areas would be easily identified. I do not think dirty bombs are practically effective, even if used by terrorists. Rather, the name "dirty bomb" may have more impact on people as a threat than its actual effect, the impact of which would be useful for terrorists. Therefore unnecessary radioactivity allergy works negatively.

Item (4), an attack on NPP, is the main subject of this section. The latter "4" of "4×4" means land, sea, sky and insider(s), and therefore an attack on an NPP also can be classified into four patterns.

Land, sea, and sky are easily conceivable as terrorist attack routes. According to experts, however, terrorism supported by insider(s) requires the highest caution. For example, an insider is like the maidservant Omatsu, in the novel "Onihei Hanka Cho" (Devil Heizo's Investigation Note) by Ikenami Shotaro, who secretly opens the locked back door beforehand for intruders waiting outside. Therefore in future, the identity of each NPP worker may be strictly checked as an antiterrorism measure, although this is regrettable.

This is a rough outline and all I know regarding nuclear terrorism.

After the Sept. 11 terrorist attacks, what put NPPs on alert most were NPP observation tours, which an unspecified number of the general public could join. Since the attacks, NPP security has been remarkably enhanced with plant observation tours virtually impossible today.

Those wanting to visit an NPP must clarify their name and address to obtain admission beforehand. Before entering the plant, they must show their photo on a passport or driving license to identify themselves. If they want to enter an access-controlled area of the plant, they must be accompanied by plant staff. If they want to enter a controlled area, their identity is rechecked before they are allowed to enter with a radiation measuring device. Since all these areas are barricaded at their entrances, terrorism via an observation tour visitor would be impossible. Even if terrorists tried to implement their objectives, they would be overpowered by security guards instantly.

The possible routes of an armed attack include land, sea and sky. Security measures against an armed attack probably differ depending on the route but remain undisclosed, so all we can do is observe from outside and guess.

The general public can only observe the security situation against a land attack around the NPP main gate. We can see strict alerts around the gate. The road leading to the main gate is zigzag and adequate counter attacks against terrorists riding vehicles such as trucks would be possible. With regard to an attack from the sea, a Japan Coast Guard naval vessel is always patrolling offshore. With regard to an aerial attack, a kamikaze attack on an NPP by an aircraft would be possible at worst. I heard a report saying that ground-to-air missiles might be deployed for a reprocessing plant in France.

Currently the scope of terrorism scenarios envisaged includes armed attacks on NPP, and wars are ruled out. Terrorism and wars differ in scale and nature. In the event of a war, repeated naval bombardments pinpointing a reactor, for example, would destroy it even though it is surrounded by thick shielding concrete. A bunker-busting bomb would be more effective, only one shot of which would crush a reactor. Conversely, all terrorists can do is engage in secretive surprise arracks using low-grade weapons at most. Therefore their handling is quite different.

6.5.2   Measures Against Terrorism by Land and Sea Based on the Design Basis Threat (DBT)

A textbook related to counterterrorism measures shows the following measures. First, conceivable types of terrorist attacks are assumed for terrorism via each of land, sea and sky. Among these assumptions, the maximum assumed threat is called a design basis threat (DBT), against which defense system and equipment as well as personnel positioning are prepared beforehand. DBT resembles the design basis accident (DBA) assumed in safety design. DBAs can be automatically coped with via plant safety facilities. Similarly, attacks up to the DBT level can be immediately coped with via defense arrangement.

The problem is an attack exceeding DBT, as in the case of a severe accident exceeding DBA. Against such terrorism, suppression by the Self Defense Forces is the only solution. Key points of antiterrorism measures are the degree to which an NPP could be damaged by terrorism and the level of nuclear disaster this could cause.

To what degree can a terrorist attack damage an NPP? The key is insider(s). The success of a terrorist attack depends on the ability of insiders and the speed of attack. Conversely, the extent to which the damage can be prevented depends on how soon the attack can be suppressed. In other words, this is a competition for speed between terrorists and a garrison force. The shorter the distance between a garrison force base and an NPP, the quicker terrorism can be suppressed. Thus an NPP far from its garrison force base must further strengthen its counterterrorism measures.

Conversely, in the eyes of terrorists, old NPPs where the same safety devices tend to be laid out side by side are favorable because they can effectively and instantly destroy such NPPs. Previously, the concept that installing the same equipment and devices in the same area was convenient for maintenance predominated. However, this has an adverse effect in the face of terrorism. The same safety equipment should be placed in different areas. This is the same problem caused by the tsunami that attacked the Fukushima Daiichi station, which rendered the emergency power generators placed in the same basement area simultaneously unusable. This is the common feature of the concept of counterterrorism measures and that of disaster prevention measures against natural phenomena.

Reflecting on the JCO criticality accident in 1999, when there was no emergency operation facility (off-site center) to instruct and communicate with the accident site, the Government spent 100 billion yen constructing off-site centers in the vicinity of major nuclear facilities. During the Fukushima accident, the off-site center in Okuma Town became unusable due to earthquake destruction and blackout. The off-site center officials, who should have coped with the accident at the center, immediately moved to Fukushima City under the pretext of increased radiation level at the center, which was frowned on by people. Both the off-site center and its personnel did not function at all and were useless.

The Government is said to be currently considering abolishing off-site centers and constructing new emergency response facilities equipped with seismically isolated structures and radiation protection measures in areas relatively far from each NPP site, but if this is true, such construction is a waste of taxpayers' money. During the Fukushima accident, the seismic isolated building in the NPP site was relatively effective as an accident response facility. Conversely, the official residence of the Prime Minister, which did not suffer from the earthquake or tsunami and has sufficient equipment and facilities, simply hindered accident responses and was totally useless. If the new emergency response facilities, each of which is scheduled to be built far away from each NPP site, are constructed, public officials knowing nothing of local NPP sites can do nothing but a waste of national coffers.

The off-site centers scheduled to be dismantled can be utilized as antiterrorism facilities instead. During normal times, they could be used as education and training facilities for some members dispatched from the SDF on a rotating basis. The very presence of the SDF personnel in the vicinity of NPP would help deter terrorism. To enhance the deterrent, merely hoisting the national flag in the facilities and holding parades of fully armed SDF personnel during commuter rush hours would be enough.

Active participation in events such as festivals would also be a strong deterrent against terrorism. Psychological pressure is quite effective against terrorists.

Counterterrorism measures should start from such routine activities.

6.5.3 Terrorism Using Aircraft

Once I conducted research into aircraft terrorism against an NPP after the Sept. 11 terrorist attacks, in which two hijacked Boeing 767 s crashed into the N.Y. World Trade Center (WTC). This research was requested by the Denki Shimbun (a newspaper publisher) and the results were posted in its newspaper under the title of Repercussions of Terrorism in the U.S. (December 3–7, 2001). Although this topic deviates from the objective of this book, aircraft terrorism is a very popular topic, so I am going to outline the results of my research.

According to a report, two of the four hijacked passenger aircraft crashed into the WTC buildings, another crashed into the Pentagon in Washington, and the final plane crashed to the ground following a scuffle between passengers and the terrorists. Later, a report said that Al-Qaeda initially planned to crash the last one into an NPP.

The WTC had about 530-m seaside buildings, from which a gigantic fireball rose after each crash of the aircraft, and a conflagration occurred, killing about 3,000 people including 24 Japanese.

Even before that, the crash of an aircraft into a nuclear facility was a safety concern. Because aircraft accidents tend to occur during takeoff and landing, safety rules were in place prohibiting the construction of airports in the vicinity of an NPP and flying above it. However, mass media ignored these safety rules without hesitation every time a trouble took place at an NPP, repeatedly violating them by flying helicopters closely above it.

Such a problem became a serious concern when the safety review of the Rokkasho Nuclear Fuel Reprocessing Facility in the village of Rokkasho, Aomori prefecture started. Because the reprocessing facility is near the U.S. Air Force Misawa Base, the crashing of an U.S. fighter jet into the facility was a safety concern. The reprocessing facility is a building with a wide roof, under which gigantic plants are standing contiguously, which is quite different from NPPs, where important equipment and components are contained in relatively small containment vessels. The reprocessing facility has a large area into which an aircraft can easily crash. The concern that a Phantom fighter may crash into the large wide roof was understandable.

Japan Nuclear Fuel Limited., the owner of the reprocessing facility, requested that the Sandia National Laboratory of the U.S. implement a Phantom fighter collision experiment. The Sandia National Laboratory in New Mexico is famous for making the first A-bomb test a success in cooperation with the Los Alamos National Laboratory.

In the experiment, a Phantom fighter weighing about 20 t was loaded on rails, accelerated up to 200 m/s with a rocket propulsion system, and crashed into a

Fig. 6.1 Phantom fighter collision experiment at the Sandia National Laboratory (Sandia National Laboratory website)

concrete wall of 3.7 m in thickness (Fig. 6.1). I watched this experiment on TV. The Phantom fighter was crashed into the wall and crumpled from its nose as if an accordion-like paper lantern had folded up. Reportedly, it took about 0.1 s to crash the fighter, which seems like a short time, but compared to an explosion or impact phenomenon, is relatively long. Conversely, the concrete wall, into which the fighter crashed, was dented by about 70 cm.

After the Sept. 11 terrorist attacks, these experiment results were introduced to related parties and highly evaluated internationally. I had thought the heavy engine of an aircraft would have a fatal destructive force, which proved wrong. The experiment showed that the entire soft body acted as a destructive force. It is said that the fuel reprocessing plant in the village of Rokkasho, Aomori prefecture, was designed based on these experiment results.

Aircraft terrorism issues posed by the Sept. 11 terrorist attacks are the following three points:

(1) Can an aircraft crash into a containment vessel?
(2) Will an aircraft crash damage a containment vessel?
(3) Will the fuel of the crushed aircraft cause a fire in a containment vessel?

I am going to answer these one by one.

The speed of the aircraft that crashed into the trade center building was about half their cruising speed of 250 m/s. One might think that crashing them with higher speed would have been more effective to destroy the buildings, which is not the case. To crash an aircraft into a target on the ground with certainty, it needs to decelerate.

The hijacked aircraft are estimated to have started descending when at least dozens of kilometers from the trade center. Although the trade center buildings were as much as 530 m high, they were very low targets for an aircraft. The aircraft are said to have crashed into the trade center buildings at a height of about 300 m, which were the higher part of the buildings. It is said that to lower an aircraft to this height, the same degree of deceleration required for landing is necessary.

I asked an expert "Why didn't the aircraft temporarily decelerate to descend and then accelerate again while targeting each building?" He laughed off my question. Reaccelerating an aircraft elevates it. Such reacceleration is tantamount to an act of canceling the target setting performed from afar. The expert told me to imagine the landing scene of an aircraft slowly decelerating while maintaining a certain angle well before landing on the runway. It is said that the angle of a descending aircraft before landing is about 3°.

This shows even crashing an aircraft into a target like the World Trade Center requires a great deal of skill. Targeting a reactor containment vessel at a height of 50 m from the ground would require a skill exceeding that. I have heard that because a turbine building, stacks, tanks, meteorological tower and various other facilities surround a reactor, crashing an aircraft into it by ducking them would require a lot of hands-on training. It is quite different from landing an aircraft onto a long runway. It is said that a skilled pilot might succeed it but would have second thought before doing it.

Crashing an aircraft into a containment vessel seems very difficult.

The next subject is whether crashing an aircraft into a containment vessel, even if were to succeed, could damage it. The Phantom fighter subject to the experiment in the Sandia National Laboratory dented the concrete wall by about 70 cm. A BWR containment vessel is surrounded by many walls of the reactor building, the thickness of which totals 4 m, thus would remain undamaged after the airplane crash.

The thickness of the wall of a PWR concrete containment vessel is about 1.3 m and that of the concrete surrounding a PWR steel containment vessel is about 1 m, and therefore their resistance to a plane crash is inconclusive.

Although there are uncertain factors such as the difference between a passenger aircraft and a Phantom fighter and collision speed, the expert who has examined this issue said destroying a containment vessel using an airplane is impossible. Because he/she declined to disclose the calculation results because of terrorism-related confidentiality, I speculated on this issue by myself and noticed a blind spot in the calculation results: one hint of which is a billiard ball.

The experiment at the Sandia National Laboratory was a crash against a flat concrete wall. A PWR containment vessel has a round shape. As with a billiard ball, if an airplane has failed to hit the center of a containment vessel, its collision energy does not act on it entirely, unlike a crash against a flat wall. An airplane having failed to hit the center core of a containment vessel would bounce away in another direction. Hitting a containment vessel, which is a low structure, with an airplane would be very difficult. Even if an airplane had hit it, it could hardly hit its center core. In terms of probability, its design safety has no problem. As in the case of terrorism by land and sea, such an attempt can be handled with post-terrorism MISSAD.

The remaining issue is a fire due to aircraft fuel. As I have discussed, it is apparent that an aircraft cannot crash into a containment vessel. A fire due to aircraft fuel, if at all, would occur outside a containment vessel. According to a report, an Airbus A300 just after taking off crashed into a residential area in New York in November 13, 2001 but the fire due to the crash was extinguished without spreading. A fire caused by a common aircraft crash does not seem a serious problem for a containment vessel.

An issue raised immediately after the Sept. 11 terrorist attacks was a scenario where a passenger aircraft crashes into the steel head of a PWR containment and its fuel, having flowed into the containment vessel, vaporizes and generates a fireball resembling those observed in the WTC. The nature of such fireball remains unknown, but judging from the television images, the fireballs from the WTC seemed different from explosion phenomena.

Some PWR containment vessels made of steel have no cover on their top and their circumference is protected only by a concrete wall (Fig. 6.2). If a passenger aircraft crashed to the top of a containment vessel, aircraft fuel might flow into it and generate a fireball. In this case, the aircraft would have to crash into the containment vessel without touching the surrounding concrete wall. See Fig. 6.2. A crash angle of at least 12° would be necessary.

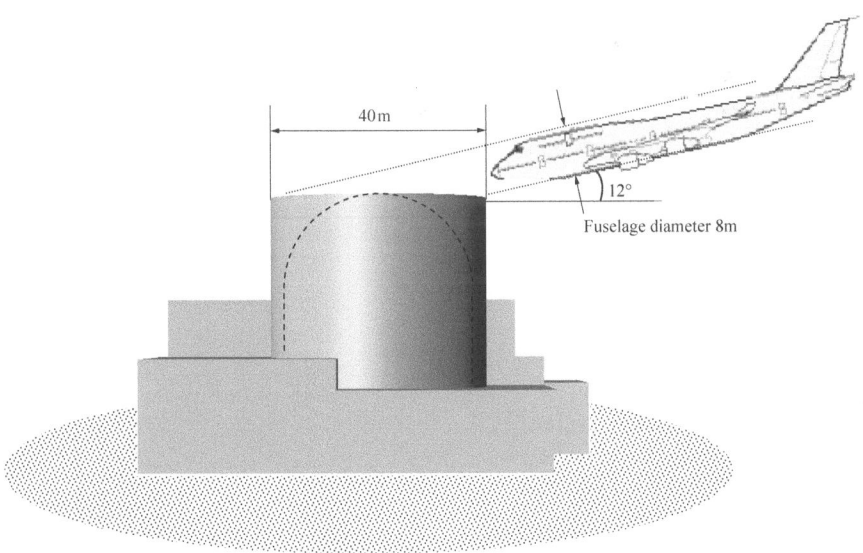

Fig. 6.2 Estimated aircraft crash angle

A descent angle of 12° is almost the same as the landing angle of a space shuttle and a nose dive far steeper than the descending angle of passenger aircraft. Would such a stunt be possible for an ordinary pilot? Moreover, the target is the center of a small circle about 40 m in diameter. The potential for such a nose dive succeeding is not zero but would be almost impossible.

I asked a skilled pilot to ask about this to an expert. The expert indirectly answered that a skilled pilot capable of stunt flying may be able to perform such nose dive crash using a small propeller plane. It is a nose dive from directly above a containment vessel using the spiral dive technique of the Imperial Japan Navy, according to him. The expert added, however, that the pilot would need time to check the target by circling the containment vessel before nose diving and that a rifle would be sufficient to shoot down such a dubious plane circling a containment vessel.

These are the results of my studying aircraft terrorism. Because NPP's containment vessel is too small to be a terrorism target, even if an aircraft crash were to succeed, causing a fire by crashing into the containment vessel as in the case of the WTC would be impossible.

Terrorism by land or sea is an armed attack on an NPP by stealth, measures against which are already in place. Once design basis threats are determined, military force exceeding such threats will repel terrorism, which corresponds to eliminating an initiating event in the case of design. The problem is the duration of terrorism. If terrorism lasts for a short time and damage is small, that means its failure. Once terrorism is conquered, all that should be done is to use the surviving systems and equipment to ensure the safety of the reactors. To make them survive terrorism, ensuring a dispersed layout of equivalent systems and equipment is effective so that they will not be destroyed simultaneously.

Conversely, aircraft terrorism is an attempt to hit the target with a single crash. Even if an aircraft crash were to succeed, the terrorists on board would perish simultaneously and there would therefore be no second attack. It is an all or nothing one-shot deal. Countermeasures and defense methods differ according to terrorism by land or sea and by sky. That concludes the topic of terrorism.

Chapter 7
Road to Decommissioning

It is said that the decommissioning of the Fukushima Daiichi NPS is scheduled for completion in 40 years starting from the investigation of the status and the whereabouts of the molten fuel by the International Research Institute for Nuclear Decommissioning established in August 2013. Although the decommissioning remains a long way off, I am often asked by mass media people what types of robots will be used and how the molten fuel will be disposed of. However, no experts having been engaged in past decommissioning work consider this Fukushima decommissioning plan will be completed as scheduled.

First, I explain what decommissioning is. Decommissioning is generally perceived as special, dangerous work because of radioactivity in nuclear facilities. Conversely, simply removing radioactive materials completely from nuclear facilities makes them ordinary facilities. Decommissioning, therefore, means work to remove radioactive materials from nuclear facilities, hence the need to precisely identify their whereabouts and plan their removal.

More than 99 % of the radioactive materials of a reactor are in its fuel rods, which is easily understandable given that nuclear fission reactions occur in uranium dioxide pellets in fuel rods. The remaining 1 % of radioactivity is attributed to the elements of materials of reactor-surrounding structures changed into radioactive elements (radioactivation) due to neutron irradiation and radioactivated impurities in water.

During normal decommissioning, fuel assemblies are removed from the reactor before launching decommissioning work. Removing fuel assemblies from a reactor is relatively easy because the method used is the same as that for regular refueling. An NPP is fully equipped with remote-handling equipment for refueling. During normal decommissioning, this equipment is used to extract fuel assemblies containing 99 % or more of radioactive materials beforehand, whereupon the remaining 1 % of radioactive materials in reactor-surrounding structures are removed. Accordingly, the amount of radioactive materials to be removed in decommissioning is relatively small.

© Springer Japan 2015
M. Ishikawa, *A Study of the Fukushima Daiichi Nuclear Accident Process*,
DOI 10.1007/978-4-431-55543-8_7

However, the decommissioning of the Fukushima Daiichi NPS is different. Fuel rods that account for 99 % of the radioactive materials in the reactor core remain in a melted and collapsed state inside the reactor. Even the precise whereabouts of the molten fuel remain unknown. Until the whereabouts of molten fuel are precisely determined and its physical properties clarified, specific planning is impossible. Developing remotely-controlled robots seems to have been proposed but a work plan should be developed first. Nothing has been scheduled regarding what kinds of robots will be manufactured and where. The mass media seem too hasty at this point.

The Government seems hopeful that the decommissioning of the Fukushima Daiichi NPS will be achieved in 40 years. I also hope so but it is probably impossible. This is because the only example of decommissioning of core melt reactor facilities completed to date is the Stationary Low-Power Reactor Number One (SL-1) reactor in the U.S., where a reactivity accident took place, and all other core melt reactor facilities remain under strict surveillance.

7.1   The Current Situations of Reactors That Suffered Core Melt

The first reactor that suffered a core melt was the Windscale graphite-type reactor in Sellafield, England. In October 1957, its reactor core melted and radioactive iodine contaminated the surrounding environment, which stopped milk being shipped from pastures in the vicinity for 8 weeks. Although the Windscale was a graphite-type reactor with a small power output, decommissioning a reactor even on this scale is difficult, and therefore its decommissioning, which started in around 1980, was stopped. This shows how the radiation level of molten fuel is high and difficult to handle.

In the TMI Unit 2, where a core melt occurred in 1979, after the molten fuel completely solidified, it was crushed into pieces before being removed. The destination for its final disposal remains undecided and it has been temporarily stored in a national laboratory in the vast desert in Idaho. Even though the molten fuel was removed from the reactor, it doesn't mean perfect decontamination. The level of contamination inside the facilities remains high and even observation tours of the site are impossible without special permission.

In 1986, the Chernobyl accident occurred in the former Soviet Union. The reactor core graphite burned completely, and all the fuel rods inside the core melted. The molten fuel fused the concrete as thick as 3 m supporting the reactor core, which dropped onto the hallway of the first basement about 2 m below, and piled up as a heap of debris. Because decay heat was generated in this mound, the center became very hot, remelted, whereupon the molten fuel broke the outer surface of the mound and flowed out of it three times, according to a report. Relatively low in viscosity, this molten material is said to have flowed 50 m on the basement corridor and finally coagulated. In other words, the molten fuel that flowed out lost heat and coagulated in a thin form like chewing gum.

In the case of the Chernobyl accident, decommissioning was ruled out from the beginning. Residents living within 30 km of the NPP were forcibly evacuated. The reactor building was covered with a "sarcophagus" and remains under surveillance. Although the situation remains unchanged, the sarcophagus has become brittle after being exposed to winds and snow for about 30 years since the accident and therefore construction to replace it is ongoing.

Although the name "sarcophagus" conjures up an image of a robust structure, in fact it's a simple barrack whose roof is an iron plate mounted on the reactor building walls and reinforced with iron pipes. The sarcophagus is neither windtight nor raintight. Although the accident occurred in the former Soviet Union, the NPP is currently owned by post-independence Ukraine. The Ukraine government does not seem willing to lay its hand on the solidified molten core, as reflected in the construction of the new sarcophagus, which is supposedly capable of maintaining safety for a century. Incidentally, it is said that, within the 30 km compulsory evacuation area, many farmers have returned to their homes and are regaining their lives. In addition, it is said that wild animals have been increasing in areas from where people had disappeared. I heard that even wolves have appeared. According to news, similar situations seems to begin to appear in evacuated areas in Fukushima.

SL-1, the only reactor dismantled and removed among those that suffered core melt, was a small reactor used to train nuclear submarine crew members. Trainees had returned to their homes during Christmas vacation, during which its maintenance was completed with fuel assemblies replaced and it was standing by for operation. The accident occurred in 1961 when a worker withdrew a control rod from the reactor core center with his hands, which triggered a reactivity accident resulting in a core melt.

The aluminum-clad metallic uranium fuel resembling a thin plate instantly vaporized, reacted with the water in its vicinity and caused a steam explosion. The fuel melted and evaporated, almost all of which scattered shapelessly in the reactor building. Water hammer force triggered by the steam explosion made the reactor pressure vessel vertically jump with its pipes torn apart and collide with the crane above it, before dropping to its original position. The jump of the pressure vessel broke and caused the collapse of the heat insulator surrounding it and the pressure vessel fell down onto the collapsed heat insulator like its cushion.

I remember that the resulting radioactive contamination was relatively minor because the molten fuel rods had been newly replaced. Although a radioactivity accident is caused by a rapid and large power output, the amount of nuclear fission is relatively small since the duration of heat generation is as short as a few milliseconds. Thus the amount of the reactor core radioactivity was trivial. The SL-1 reactor building was decontaminated and then partially used as a research laboratory. During my stay in the U.S. as a junior researcher at JAERI, I spent about 2 weeks in the laboratory.

During the nuclear reactor development period in the 1950s, when Japan began to restart atomic research after having been prohibited from conducting it because of its defeated nation status, many criticality accidents (reactivity accidents) took place around the world. Back then, the mechanism of the nuclear fission chain reaction itself was not well known, and there were no computers yet, which are quite helpful today. In those days, experiments were performed with radiation measurement devices installed at various locations of a criticality experiment device, while observing the measured values, using hand calculations and intuition, and by loading fuel and withdrawing control rods. Such experimental facilities were too simple to be called reactors and were collectively referred to as criticality facilities.

Reportedly, about 60 criticality accidents were caused by criticality facilities at the time. Most of these facilities are unusable today and I think many were disposed of (decommissioned). Such disposal was too small in scale to be called decommissioning because the nuclear fission involved was small. They probably manually removed the disposal measuring worker's exposure doses.

During the initial nuclear energy development period, a considerable number of research reactors, including those for military purpose, were used. They were designed variously and many of them were surrounded with an unnecessarily thick and robust shield because of immature radiation protection technology at the time. Needless to say, they are no longer usable and were disposed of as "white elephants." With regard to their decommissioning, some were dismantled and removed and others entombed (permanently disposed of) after removing their fuel rods. Already 60 years have elapsed since nuclear energy development started, which means most of their radioactivity due to activation will have attenuated or disappeared and probably enabled their burial disposal.

7.2 Decommissioning in the Earliest Days and Today

The dismantling and removal of commercial power reactors started in the mid-1980s. The Shippingport Atomic Power Station (SAPS), equipped with as many as three containment vessels in the U.S. mentioned previously, the graphite Windscale Advanced Gas-cooled Reactor (WAGR) in England, and the Japan Power Demonstration Reactor (JPDR) were the vanguards. Interestingly, the methods of dismantling and removing reactors used in the early days had different characteristics, reflecting the state of affairs in each country.

The decommissioning of the SAPS, which was the earliest among the three, was performed by removing and sealing large radioactive equipment such as the reactor pressure vessel and steam generator with their shape maintained, housing small radioactive equipment in the pressure vessel and then solidifying them with concrete. The dismantled equipment was then transported to the Hanford Site for waste disposal.

The SAPS is located in the vicinity of the junction of the Allegheny and Ohio Rivers near Pittsburgh; both of which join the Mississippi River, which flows into the Gulf of Mexico. The SAPS located inland of the eastern U.S. were very far from the Hanford Site (waste disposal facility) located in Washington on the U.S. West

Coast. The land transportation of radioactive waste generated from decommissioning such as the reactor vessels and steam generator required a huge cost. Moreover, the period was immediately after the TMI and Chernobyl accidents, hence the transportation of radioactive waste was a sensitive issue. The U.S. government had second thoughts about transporting it by land.

However, parties engaged in the decommissioning demonstrated a frontier spirit by transporting the radioactive waste on a barge through rivers, like the story of The Adventures of Tom Sawyer. They achieved the transportation using a flat-bottomed barge by negotiating the Mississippi River, sailing across the Gulf of Mexico, notorious for its typhoons, and through the Panama Canal, going northward to the high-wave Pacific Ocean and then to the Hanford Site located near the middle Columbia River.

Since the barge was flat-bottomed, navigation on high-wave seas was unstable. Had it encountered a typhoon as is often the case in the Gulf of Mexico, it would have sunk immediately. However, they had checked the weather beforehand, and achieved the transportation bravely and successfully, showing their Yankee spirit. Today, the transportation of radioactive materials by land is performed without a problem.

Conversely, England had announced a decommissioning plan spanning up to 17 years from the very beginning, which was intended as a measure to counter unemployment among the NPP staff. There was a scenario where most the NPP staff members would reach retirement age within 17 years and learn the knack of decommissioning work. In the meantime, radioactivity would attenuate to some extent and a remotely-controlled machine or similar suitable for decommissioning might be developed by neighboring countries someday. It was quite a leisurely mindset. It is said that carpenters in England buy an old house, live there, remodel its interior and then sell it as a new house whereby making a living. I think England is traditionally a laid back country.

One case of decommissioning in Canada, which has a large territory but a relatively small population, was also unique. For example, machines and equipment in nonradioactive large buildings such as turbine building were removed first, whereupon the large buildings were remodeled and used as malls and office buildings. Radioactive materials removed during decommissioning were collected and stored in the reactor building. Because most of the radioactivity of materials such as cobalt and iron was expected to decay during their storage in the reactor building for a century, the dismantling and removal of the reactor building were scheduled to start a century later. It was presumed that the amount of radioactivity would decrease sufficiently for dismantling after a century had passed but regrettably the dismantling plan remains unfinished. This dismantling method is so-called "mothballing." This method, however, is said to be relatively similar to the immediate dismantling and removal method in cost terms, given its long-term management expenses.

Conversely, Japanese people are diligent and hasty in general. Because of the restricted land area of Japan, decommissioning plans also tend to envisage the post-decommissioning use of reactor sites.

The decommissioning of the Japan Power Demonstration Reactor (JPDR) of the former Japan Atomic Energy Research Institute (JAERI), where I was once engaged, was an experiment in such post-decommissioning use of reactor sites. The JAERI developed a decommissioning plan; not only for the dismantling and removal of the JPDR but also for various other research. For example, the volume of radioactive waste generated during the decommissioning process was minimized and the performance and usability of various cutting machines were compared. The large reactor pressure vessel was cut into pieces using remotely-controlled cutting machines, which were then compactly filled into containers. These reflect the diligent characters of Japanese. Various cutting machines, dismantling machines and containers were developed and used experimentally. These achievements were highly evaluated in foreign countries. Looking back, these were luxurious experiments only a research institute could afford.

As shown above, initial decommissioning methods differed among countries depending on their state of affairs.

Today, many decommissioning activities are performed in a way resembling ordinary dismantling work and utilizing the above experiences. While only highly radioactive sections are dismantled using remotely-controlled machines and equipment, other sections are dismantled manually using ordinary dismantling tools, which is quicker and reduces radiation exposure more effectively. Currently, normal decommissioning work is almost the same as ordinary dismantling work, excluding special aspects whereby the former requires work in highly radioactive areas and the elimination of radioactivity. Naturally, radioactive materials are completely removed from each reactor site.

Because decommissioning technology has progressed to this extent, in the decommissioning of an NPP which has stopped operation under normal conditions, in other words during normal NPP decommissioning, the focus has shifted to utilizing post-decommissioning reactor sites and invigorating the local community. In general, an NPP is constructed in a low-population area, and as the construction progresses and the NPP starts operation, people gather around it and develop the area into a town. Today in a decommissioning plan, the redevelopment of a decommissioned site is planned beforehand to utilize the vast land for residents living nearby. Although many people may have a negative image of decommissioning, focusing only on the dark aspects such as the removal, processing and disposal of radioactive waste, such an image is outdated. For example, in the same way as the Mitsubishi Heavy Industries, Ltd.'s former Yokohama shipbuilding yard constructed more than a century ago was redeveloped as Minato Mirai 21 and became the center of Yokohama City, the redevelopment of a post-decommissioning NPP site is possible. Actually, such examples exist abroad, where after nuclear facilities were dismantled and removed, research facilities unrelated to nuclear power generation were built.

7.3   Advice to Fukushima

In normal decommissioning, immediate NPP dismantling is possible and the site is reusable, but this is not the case for the Fukushima Daiichi NPS, where fuel rods accounting for more than 99 % of the in-core radioactivity melted and their whereabouts have yet to be clarified. Planning the NPS decommissioning initially requires their whereabouts to be investigated.

For this, it is crucial to install measurement equipment on the refueling floor (the 5th floor of the reactor building), directly above each reactor, to prepare a molten fuel distribution map based on detailed measurements. After grasping the whereabouts of most molten fuel, its properties must be investigated by its sampling. The investigation should also clarify the physicochemical properties indispensable for decommissioning such as whether the molten fuel is hard or soft and whether it is readily soluble in water. Only after such properties are clarified will specific decommissioning planning be possible. Full-scale decommissioning work comes much later.

Under current circumstances, even installing measurement equipment directly above each reactor is not easy. This is because of the high radiation dose and the fact that stacked debris contaminated by explosions is hindering workers' operations. In the first place, even the plan of measuring the molten fuel under the high-radiation environment has not been developed.

TEPCO's engineers have sufficient spirit to bravely cope with this work, but key to the success of their efforts are financial resources, time and warm support. Financial resources and time can be managed in any way in today's Japan but the problem is people's warm support. Japanese people are sympathetic with the disaster victims but cold and harsh against TEPCO, the owner of the accident nuclear power plant.

However, other than TEPCO's personnel knowing the Fukushima Daiichi NPS best, are there any candidates expected to be engaged in dangerous decommissioning? Even if there were any, would they have the knowledge of the equipment and nuclear power plant machines? To decommission the NPP, the nuclear-related knowledge of TEPCO is indispensable. Whether TEPCO's staff can work actively and positively decides the issue of the decommissioning of the Fukushima Daiichi NPS.

Three years after the accident, residents who once lived near the Fukushima site may still have hard feelings against the accident. However, I hope they will change their hard feelings into hope for future recovery and redevelopment. Decommissioning should be regarded as work leading to the future development of the disaster area, not simply as the processing and disposal of radioactive waste. TEPCO needs the warm support of local residents.

In the TMI NPP, most of the molten core was removed but its facilities still have high radioactivity, meaning entry and exit there without permission remain prohibited. The Chernobyl NPP declared that the accident had been brought under control but the molten fuel was left almost unattended. These two NPPs adopted diametrically opposing approaches to the handling of the molten fuel: crushing the solidified molten fuel and removing it or leaving it as is. As for the decommissioning of the Fukushima Daiichi NPS, either crushing solidified molten fuel and removing it, or storing and managing it in a state harmless to the surrounding area, will be possible. The option is in the hands of the Government.

The problem is the extended period required for both options. How the molten fuel should be managed during such period is also a key point.

The TMI NPP launched a study under international cooperation to investigate the situation of the molten fuel (Fig. 1.1). Access to the Chernobyl NPP was also granted to visitors from abroad one year after the accident, since that time access to the post-accident site has been relatively free. It is high time Japan learned from these examples.

Foreign countries interested in the post-accident situation in Fukushima are getting tired of awaiting related information from Japan, which has failed to transmit it to the world. They are indignant because the facts of the accidents such as the process of the core melts and explosions remain undisclosed and the problem of contaminated water has surfaced only recently. They are growing impatient, wondering "What is going on? How can we help them when they disclose nothing?"

Foreign countries have no idea what Japan is going to do. Japan's policy of not restarting the idling NPPs in the country for the time being while the Prime Minister and manufacturers are actively marketing nuclear reactors abroad is contradictory. I am concerned that a statement by General MacArthur immediately after WWII describing the mental age of Japanese as 12 years old may resurface.

What should Japan do now? The answer is simple: learn lessons from precedents. Although the smooth disposal of highly radioactive materials in contaminated areas is difficult, such areas can become an invaluable experimental field for researchers. Land and seashore areas where radioactivity remains high and residents cannot return could be partially classified as special zones earmarked for an international research institute. In the special zone, studies concerning radioactivity, not limited to nuclear power generation but covering all radioactive phenomena, could be performed freely while seeking research themes, such as space station, from abroad.

Research subjects could be topics of interest for local residents, such as "Why are wild boars increasing despite the high radioactivity? How about their cancer?" and "How is the health condition of fish whose internal dose is 1,000 times more than normal?" The special zone could be a place where any radiation-related research is possible. Nuclear allergy peculiar to Japanese would be gradually healed if the effects of radiation on the human body and agricultural products, for example, are specifically analyzed and verified and the results presented to the public. Such efforts have been shunned to date.

Decommissioning by TEPCO could be performed in cooperation with the above-mentioned international research institute, which would create a sense of camaraderie and a further motivation to decommissioning work because of new human relationships, abundant topics and stimulation. Moreover, researchers from overseas would communicate the situation and significance of decommissioning in the post-accident site to the people of their countries. Decommissioning could progress incorporating the result of research institute activities.

The participation of overseas researchers in the Japan's post-accident handling would automatically dissipate the dissatisfaction. Moreover, their experiences in Fukushima would also be globally disseminated, enhancing understanding and trust toward Japan.

Constructing a heliport in the special zone would facilitate access to the inconvenient Fukushima site, which would boost the number of visitors from abroad and create an international community.

Although the fact that the accident occurred in Fukushima and contaminated it is very sad, experiences and research findings in the very high radiation fields caused by the molten fuel should be useful somewhere worldwide provided the peaceful use of nuclear energy continues. Exploiting the current situation, rather than mourning it, will help resolve the future of Fukushima.

Chapter 8
Outcome of the Study

After studying each process of the core melts and hydrogen explosions, I have presented my views regarding the Fukushima accident including radioactive contamination and evacuation, station blackout and the construction of the seawall, the reconstruction of nuclear safety, and decommissioning. In this chapter, I am going to summarize the major results of my studies on these subjects.

First, the cause of the core melt was not decay heat, but was a chemical reaction between the fuel cladding material zirconium and water.

The major difference between the Fukushima Daiichi and TMI accidents was only the amount of injected water into high temperature cores. In the TMI NPP, injection of a large amount of water instantly triggered the collapse of the reactor core, and subsequent core melt. In the Fukushima Daiichi NPS, the amount of water injected from fire-extinguishing pump was limited, and resulted in a time lag between the core collapse and melt. Different events in each units have made the overall accident situation significantly complicating.

However, the results were the same for both accidents: the core melts occurred after the emergence of a large volume of water. In both cases, a large amount of hydrogen gas was produced by the reduction process in short-lasting chemical reactions, which caused explosions. I have spent most of this book in discussing the identical nature of the chemical reactions.

The second point is the route of hydrogen gas that leaked into each reactor building.

The pressure and heat caused by rapid hydrogen generation in the reactor core lifted up the containment vessel head, creating clearances, from which the hydrogen gas blew out, further lifting the shield plug located above the containment vessel head and causing significant leakage into the 5th floor (refueling floor). Other than by this route, the explosion at Unit 1 cannot be explained reasonably and the significant amount of hydrogen gas capable of explosion could not have leaked into the 5th floor.

© Springer Japan 2015
M. Ishikawa, *A Study of the Fukushima Daiichi Nuclear Accident Process*,
DOI 10.1007/978-4-431-55543-8_8

The ignition source of explosions at Units 1 and 3 was the impact of the dropping of the shield plug of each unit, which had been lifted by hydrogen pressure into the air and fell down, causing the impact.

I estimated this based on the example of the Chernobyl accident, where the shield plug floated in the air due to hydrogen gas pressure and somersaulted due to the subsequent explosion.

The third point is the explosion at Unit 4.

The cause of the explosion at Unit 4 was hydrogen gas generated from the molten core of Unit 3, which flowed back through a ventilation duct into Unit 4. It was an explosion in the form of a by-blow from Unit 3 and its ignition source was the breakage of the ventilation duct due to its thermal expansion.

The U.S. government initially thought this explosion was caused by water leakage at the earthquake-damaged spent fuel storage pool and was concerned about the potential for the global spread of radioactive materials. Although this concern proved groundless, water leakage from a spent fuel storage pools is being highlighted as a new safety issue.

The fourth point is the final measure to prevent core melt, namely the depressurization of and water injection into reactor.

This method is quite effective, provided depressurization is followed by immediate water injection. If water is supplied to fuel rods upon their cooling by reactor depressurization, the reaction of zircaloy with water will never happen, hence the core will not melt.

Also in the Fukushima case, the idea of water injection itself was correct but did not succeed because of the extended time required. In both Units 2 and 3, water injections were delayed by about 2 h or longer, which heated the fuel rods and triggered an oxidizing reaction between the fuel rods and the water. If water is injected earlier, core cooling is possible even though minor leakage of radioactive materials may occur.

I hope people working for NPPs bear in mind this final measure for cooling a reactor core, i.e. reactor depressurization followed by immediate water injection.

The fifth point is the high effectiveness of the SC vent.

This is the first proof of SC vent actually operated in the world. As I explained in Part II, Chap. 4, the decontamination factor of SC vent is 750 (judging from the radiation level transition during the accident), which is a high decontamination effect as high as nearly four digits.

During the time period when the vent valve was open, the background radiation level near the main gate was estimated as equivalent to an annual radiation dose of about 20 mSv. The background radiation level in the nearest residential area is estimated to be lower than this value, suggesting that nearby residents need not engage in emergency evacuation provided the SC vent works normally. There is also no need to install a filtered vent for a BWR. Rather, the focus should be placed on reinforcing and improving the efficiency of the existing SC vent.

The sixth point is the reference radiation level based on which evacuation is recommended.

The Government performed the evacuation of residents without a predetermined and specific evacuation dose. If the Government had adopted 20 mSv/y, which is the lower limit of the evacuation dose range repeatedly recommended by the ICRP, there would have been no need to forcibly evacuate residents late at night on the day of the earthquake and the confusion could have been prevented. The tragedy of the death of more than 60 residents during the evacuation could also have been prevented.

Had the Government adopted the dose of 100 mSv/y, which is the upper limit of the dose range recommended by the ICRP, areas requiring evacuation would have been far more limited and with far fewer evacuees, the infrastructure destroyed by the earthquake could have been recovered much earlier and lives in most of the Hamadori area would have continued like before.

The press criticized the Government's mistake of not having used the calculation results of the System for Prediction of Environmental Emergency Dose Information (SPEEDI). Most culpable here is the mishandling of the emergency by the Government, which did not comply with predetermined rules and systems, issued orders as it pleased and triggered failures.

The seventh point is the second cause of the accident, i.e. the station blackout.

I hope you understand that the aforementioned short-time blackout rule, which is being criticized, had been determined following in-depth discussions and examinations. Conversely, it should be noted how public opinions led by mass media are sometimes whimsical and imprudent.

I feel that it is my duty to record the fact of that timing as one of a few surviving first-generation members of the nuclear community in Japan who have been engaged in nuclear energy development.

The eighth point is the reconstruction of nuclear safety, which is most important.

I have discussed the concept and history to ensure nuclear safety that have been adopted since the start of nuclear power generation and clarified what was lacking in Japan.

What Japan lacked were not appropriate safety designs of NPPs as misunderstood by the general public but emergency measures (emergency readiness) against disaster factors in the environment surrounding each NPP. Such emergency measures are essential just like a car society requires comprehensive measures to ensure traffic safety. I explained the difference between the concept of nuclear safety and MISSAD, detailed the concept of MISSAD to be performed in future. I am happy if this concept is adopted in future NPPs.

It is apparent that constructing seawalls is both harmful and useless. If they are constructed, the cost would be added to electricity charges, which is nothing more than a waste of money.

The ninth point is the decommissioning of the Fukushima Daiichi NPS, a major topic of these days.

While decommissioning requires an extended period, nothing can be solved by a mindset focused on eliminating radiation-contaminated nuclear facilities as soon as possible. The same can be said about contaminated water, which has become another major topic today.

Part II, Chap. 7 discussed the decommissioning issue from the perspective of how the contaminated areas in Fukushima could be utilized for its reconstruction. Global mindsets in the early days of decommissioning, which I learned during my hands-on experiences in decommissioning, were very informative for me. I think you now understand after reading Chap. 7 how decommissioning methods differed depending on the national characteristics.

Chapter 9
New Insights Obtained from the Accident – To Commemorate the Publication of the English Edition

One year has passed since the Japanese version of this book was published in Japan. On the occasion of the publication of the English edition, I summarized the circumstances and events in the Fukushima accident, and further reviewed what could not be written a year ago concerning the future safety issues.

In particular, I would like to emphasize next three features as clarified through the study on the Fukushima accident.

First, the origin of core melt and a huge amount of hydrogen gas generation was not simply due to a lack of water, as commonly conceived, but due to an intense zircaloy-water reaction as a result of plenty of water supply into super hot fuel rod.

Second, judging from the careful analysis of the observed data on radioactive release, the SC vent had a very high decontamination coefficient, suggesting sufficiently high potential of improving safety concept and changing safety measures to prevent severe accident. This will be commonly applied to light water reactors world wide.

Third, I propose to develop a new concept "mitigation safety system against disaster (MISSAD)" that will be deployed the territory of the plant site. It is based on the insight that nuclear safety against risks which are beyond our common scopes of accidents could be strengthened with the society.

You may understand why I come to these points by reading the subsections which follow.

9.1 Insights and Lessons Learned from the World-First Experiences

World-first experiences in the Fukushima Daiichi NPS accident characterized by a gigantic earthquake of magnitude 9, subsequent massive tsunamis and station blackout. What we experienced with the progress of the accident include, among others,

© Springer Japan 2015
M. Ishikawa, *A Study of the Fukushima Daiichi Nuclear Accident Process*,
DOI 10.1007/978-4-431-55543-8_9

complete evaporation of cooling water from the pressure vessel, water injection from fire engines, first SC vent operation and radioactive releases, hydrogen gas spewing from the top flange of the containment vessels and resulting explosions in the reactor buildings.

The insights and lessons from these world-first experiences are summarized below.

9.1.1   Nuclear Power Plants That Withstood a Massive Earthquake of Magnitude 9

No-one in the seismological community in Japan predicted the risk of a magnitude-9 class earthquake hitting nuclear power stations. However, major facilities of 15 nuclear reactors in five nuclear power stations located on the Pacific coast in the northern part of the main island hit by the gigantic earthquake, have remained unharmed even of such a high magnitude earthquake. This indicates the adequacy of the present guide for reviewing the seismic design of nuclear power reactor facilities in Japan, though the earthquake prediction is still imperfect.[1]

9.1.2   Safety Design to Mitigate Natural Disasters Learned from Tsunami Damage

At the Fukushima Daiichi NPS site, where preparations for decommissioning is underway, tsunami damage, such as heavy oil tanks which had been uplifted and swept away, was removed, but the wreckage of a car caught between the reactor building and a thick pipe still remains.

The new regulations in Japan require the design of nuclear power stations to increase the potential tsunami height and the construction of seawalls exceeding the design height to prevent tsunami hazards. The seawall is one of the measures against a tsunami, but increasing its height is merely an input of enlarged values to the design for natural disasters as practiced previously. This cannot be an essential disaster mitigation measure.

---

[1] Possible damage to pipes in Unit 1 was implied in a report of The National Diet of Japan Fukushima Nuclear Accident Independent Investigation Commission. This implication, however, conflicts with the conclusion of the Japanese Government Investigation Committee on the Accident at the Fukushima Nuclear Power Stations, causing international repercussions. There is no existing data showing pipe rupture in Unit 1 within an hour from the occurrence of the earthquake to the onslaught of the tsunami. It is, therefore, fairly unjust or clearly wrong to assert pipe rupture only in Unit 1 out of six reactor units in Fukushima Daiichi NPS without clear evidence. The accident analysis conducted by the Nuclear Regulation Authority also denied this implication [1].

The design for structures important to safety against natural disaster has been considered sufficient, with an upper limit including a certain margin for the worst-case scenario based on past events. This concept has been common worldwide.

Only seismic designs differ from other natural disaster countermeasures. Dynamic vibration analyses for safety related structures are required using seismic waves determined from the surrounding geographical features, in addition to providing a margin for the worst case scenario. This seismic design approach proved very effective for earthquakes; all control rods in all ten operating nuclear reactors had been inserted properly and the cooling process started for shutdown even in the worst-case situation of an unprecedented magnitude-9 earthquake. The reactors in Fukushima Daiichi were in the process of cold shutdown as usual before the attack of the tsunamis.

With the progress of global warming in recent days, the scale of natural disasters tends to be far greater than before and damage tends to be much more serious. Consequently, the safety design approach for natural phenomena, which reflects the worst case as reference data, including tsunamis, must be considered obsolete. As in the seismic design, we must identify the factors (threats) triggering hazards in a scientific manner, and employ measures to cope with these factors. We have to make this effort in collaboration with countries all over the world. It may take more than a decade but the outcome would dramatically increase the robustness of nuclear power stations against natural disasters.

9.1.3 Countermeasures for Station Blackout

A station blackout occurred in the Fukushima Daiichi NPS due to tsunamis, and lasted for nearly 10 days, significantly damaging the power station. Of a total of 13 emergency generators installed in Fukushima Daiichi NPS, only one air-cooled generator for Units 5 and 6 remained intact. Many distribution panels installed on the ground floor or basement of the turbine building were submerged under water and out of operation.

Under these circumstances, everyone believed that it would have been almost impossible to use the plant facilities immediately had the power been restored. However, situation in Fukushima Daiichi NPS rapidly began improving after the installation of a temporary power supply to restore the power around March 20. This suggests that the best strategy for coping with a station blackout is to supply electricity as soon as possible, where it means to provide various types of power supply.

9.1.4 Reactor Water Level and Core Melt

In nuclear accidents, all core melts at the TMI and Units 1 to 3 in Fukushima Daiichi NPS started with cold water injections onto hot fuel rods. Prior to the Fukushima Accident it was believed that the core melt starts when a reactor water level decreases from the top of the core. However it is not true anymore. Note the reactor water levels at the time of the core melt. In the TMI accident, core melt occurred when the reactor water level dropped to about a half the core height. In the Fukushima accident, the core melt occurred in Unit 3 when the water level reached nearly the bottom of the core, and in Unit 2 when the water level that had dropped to 1 m below the bottom of the core was restored to the bottom of the core by water injection. These facts revealed that core melt is unrelated to core water level.

When the water in the pressure vessel in Unit 1 had completely lost, the core which turned into a radiant heat transmission mode gradually raised its temperature. And a part of the fuel reaching high temperature separated and fell off through the bottom of the pressure vessel to the floor of the containment vessel and piled up to form a mound. The core melt started when the heap of debris on the floor, which had become very hot internally, encountered the water.

Viewing this chain of events differently, the core of the light water reactor maintained its structure for a certain period even after the complete loss of water in the core. It was approximately a couple of hours later from the core exposure that a core had begun to collapse due to a lack in radiative cooling, as demonstrated in Unit 1. The core of the light water reactor is robust, and does not easily melt unless no water is injected into the heated core. This is a new insight obtained from the Fukushima accident. It is not correct either to have a view that the core of the light water reactor melts and liquefies.

The core of the light water reactor does not melt easily because of the high melting point of the fuel materials. Core melting and liquefaction is a figment of our imagination. This is because we only know the world of 1,000 K. This may be a new insight by the Fukushima accident.

9.1.5   Water Injection from Fire Engines

Water injection from fire engines was a desperate choice amid the unprecedented station blackout and perhaps the trickiest action in the history of nuclear accidents. However it would be useful means in emergency to cool down a nuclear reactor.

General fire engine pumps in Japan are designed to have a delivery pressure of about 0.4 MPa and a discharge rate of about 25–40 tons/h. The calories required to evaporate 40 tons of water are about 30 MWh. In other words, water injection from fire engines may be effective for cooling the reactor if the decay heat drops to around 30 MWt.

Suppose a general nuclear power station with output of 1,000 MWe. The heat output from the reactor for producing 1,000 MWe of electricity is about 3,000 MWt. The decay heat will drop to less than 1 % within 3 h of the reactor shutting down. This means water injection from a fire engine may be effective for cooling a 1,000 MWe class reactor 3 h after shutdown in terms of calories.

Incidentally, the fire engine owned by TEPCO in Fukushima Daiichi NPS is said to have a capacity twice as large as general fire engines. This fire engine was, therefore, completely capable of cooling 460 MWe class Unit 1, and 780 MWe class Units 2 and 3, respectively but in reality on the accident site, the high reactor pressure prevented water injection by fire engine pump.

9.1.6 SC Vent

In the event of an accident, the containment vessel is the most critical structure and has been regarded as the last resort of nuclear safety. Therefore it is designed and manufactured as the final barrier to confine radioactive materials.

When the vent valve is opened to reduce internal pressure and prevent damage to the containment vessel, radioactive materials confined in the containment vessel are released to the outside environment, and the local residents are exposed to radiation. Accordingly, the vent has been understood as the unfavorable step accompanying the evacuation of people.

The SC vent was actually used at the Fukushima accident, and the subsequent field data showed that the decontamination factor was 750 or more. This is a crucial fact that may largely change a concept of containment vessels in nuclear safety. Design and inspection of containment vessels, and safety measures during an accident shall be amended in near future. It is a new insight in the Fukushima accident.

On March 15, 2011, just after the accident, the background radiation level measured by monitor car near the main gate of the power station reached maximum radiation level, 1,500 mSv/y. This level of radiation was caused by direct releases of radioactive materials through the molten core in Unit 2, after the containment vessel had broken due to the failure of the SC vent. If this amount of radiation had been released through the SC vent, the radiation level would have been reduced to 1/750, which is approximately equivalent to the background radiation level of 2 mSv/y, or about one tenth of the first background radiation level. As shown in the above estimation, the significant effectiveness of the SC vent in filtering radioactive materials should be emphasized as a countermeasure for radiation release.

The SC is simple structure in which the water holds radioactive materials. Since this structure has not been fully optimized, it seems easy to modify in order to improve its decontamination efficiency, relatively in a short time. Once this is achieved, the background radiation level would be reduced to a severalfold of the normal level during ordinary operation, which would eventually eliminate the potential risk for evacuation.

9.1.7 Slow Cooling with Reactor Core Depressurization

It is not a new finding, and rather an easy knowledge of thermal hydraulics. Depressurization of reactor by venting steam provides a fuel cooling effect, which is crucial. Every nuclear expert knows this as a basic characteristic that appears in the course of the loss of coolant accident (LOCA), but it seems to have been care-lessly forgotten under the excited atmosphere at the Fukushima accident because of an excessive focus on water injection which made the core of Unit 2 and 3 melt. Details are explained in the next section.

9.2   What Could Have Been Done to Prevent the Accident?

After publishing the Japanese edition, I had many opportunities to discuss on my considerations in the book, and the audience outside the nuclear community often asked me "What can be done or could have been done if the water injection caused the core melt?" I have two answers to respond to the question. The first one is the execution of SC vent without hesitation and second is the water injection after depressurization of reactor pressure vessel without time delay.

As explained in Sect. 4.1.1 of Chap. 4, Part II, the potential decontamination fac-tor of the SC vent in the Fukushima accident is estimated as high as 750. On this base, an early vent operation is very much recommended when there is a risk of core melt. This is because if the vent valve of Unit 2 was opened, the background radia-tion level will drop to 1/750, or about 2 mSv/y.

If the vent is opened immediately after the accident, low pressures are main-tained which eliminates the risk of pressurized damage of the containment vessel. If a fire engine is available, water can be injected after depressurization of reactor, and makes it possible to continue core cooling and prevent the core melt. Eventually, it reduces the evacuation area significantly.

If seawater had been injected immediately after depressurization, the core melt in Units 2 and 3 would have been prevented (see Sect. 2.5.2 of Chap. 2, Part I).

Actually, by opening the safety relief valve, the reactor pressure was 0.5 MPa in Unit 2 and 0.4 MPa in Unit 3, and the temperature of the remaining water in the pressure vessels was supposed to range between 140 and 150 °C. Depressurization takes around 30 min, during which the fuel rods must have been cooled slowly to the saturation temperature, therefore core melt and hydrogen explosion could not have occurred.

Unfortunately, seawater injection in Unit 2 was delayed more than 2 h, and sus-pended for about 2 h in Unit 3. In both cases, the decay heat raised the core tempera-ture again, and core got melt by injecting water.

Because of its importance, I will repeat the explanation. The progression tells us that, if seawater had been injected into reactors after depressurization without delay, there would have been no time for the temperature of the fuel rods to rise again, and the cool cores would have been soaked with seawater. Accordingly, the core melt would not have occurred, and hydrogen explosions could have been prevented.

This is a new insight revealed by the Fukushima accident, and the final counter-measure to a severe accident – information that will encourage all station operators worldwide. As a matter of course, operation guides for severe accidents should be revised in future based on this fact.

As the accident data of Unit 3 suggests, cooling with the HPCI may be as effective as core depressurization, because the steam that flows into the suppression chamber via HPCI turbine. In addition, water can be injected from the fire engine provided the vent is opened. This method can also be considered as a last resort for cooling the core.

9.3   Improvements to Be Considered

Based on these new insights, let's consider improvements in the safety design of nuclear power stations.

9.3.1   Enhancement of SC Vent Performance

The SC vent system is composed of a very simple structured device. There may be a large room to improve its decontamination capability by a factor of 10.

If it were achieved, decontamination factor of the SC vent is estimated to be 10,000. If so, the background radiation level at the time of an accident would be decreased to 0.15 mSv/y even though the core melt takes place. This radiation level is not much different from those occurred during ordinary operation. Accordingly, evacuation in the event of an accident would not be necessary at least from a scientific view point.

9.3.2   Containment Vessels

The leak rate of the containment vessel has been very strictly controlled because confining radioactivity at an accident was considered crucial for maintaining safety in the past. However, if the SC vent, is such an efficient means of decontaminating radiation, it would be safer to open the SC vent to reduce the pressure of the containment vessel. Even if a core melt takes place, only a small amount of radiation may be released. Therefore, the need to strictly control the confinement capability of the containment vessel could be eliminated.

The new roles of the containment vessel will be changed to preventing direct releases of radiation during an accident and preventing minor radioactive leaks during ordinary operation. Accordingly, design requirements and operation rules for containment vessels should be changed. The evacuation plans for local residents around the nuclear power station will also be affected by these changes.

The Nuclear Regulation Authority specifies the addition of a filter vent system as a new standard, but the reduction in radiation by this system will be small if the SC vent is provided. Moreover, safety issues arising from the addition of a filter vent must also be examined.[2]

9.3.3   Reactor Buildings

The concept of the reactor building design will also be changed. To date, the reactor building for the BWR has been designed as a semi-airtight structure and referred to as the second containment vessel or the 5th radioactive defensive barrier. Consequently, hydrogen gas having leaked into the refueling room on the top floor of the reactor building turned explosive when mixed with air inside the room, caused a big explosions in Units 1 and 3.

Based on these facts, the following improvements can be suggested: When the required seismic performance and other requirements are met, the reactor building is not necessarily equipped with a strong containment capability, but can be a semi-closed structure. In extreme terms, a building containing glass windows may also be applicable. Of course, the presence of the SC vent is the prerequisite for improvements.

9.3.4   Study of Cooling by Radiation Heat

We need a study on radiation heat at very high temperature, say 1500 ~ 2500K, which is the world we are not used to in daily life. Detailed studies are required for radiation heat as a means of cooling the whole reactor components (e.g. core, pressure vessel and containment vessel) too. The decay heat will be lower to a level of around 0.5 % in 1 day. This level of heat may be removed by radiation depending on some resourceful ideas. By combining this with SC water cooling, a reasonable solution is more than feasible.

---

[2]Consider the condition immediately after emissions of hydrogen gas from the containment vessel. The filter vent installed far from the containment vessel will be filled with relatively hot hydrogen gas, at a temperature of at least 100 °C, or more. When the hydrogen gas filled in the filter vent unit cools down over time, its volume shrinks, reducing the pressure of the filter vent unit. When the pressure drops, it is likely that ambient air will flow into the filter vent unit, whereupon the hydrogen gas that encounters the air will become explosive.

In short, the filter vent is a safety device with the potential for explosion, which cannot be ignored. The new standard provided by the Nuclear Regulation Authority requires sufficient consideration of safety (hydrogen explosions in particular).

The PWR is an exception. The large containment vessel of the PWR is filled with air, and as explained in the TMI accident, hydrogen that accumulates in the containment vessel during a core melt will become an explosive gas. The potential for explosion inside the containment vessel, as happened at TMI, cannot be ignored. Accordingly, the explosive gas in the PWR must be purged outside to make it non-explosive by thinning with air, but this process requires radioactive materials to be eliminated from the hydrogen gas, for which a filter vent is used.

If, however, the water in the pressure vessel is completely lost, and radiation heat is used to cool the core, provisions should be made for a part of the core dropping onto the floor of the containment vessel depending on the strength of fuel materials at high temperature and the fusing characteristics of the core structural materials.

My friend in the U.S. e-mailed me just after the accident, saying "You should have dropped debris on the bottom of the containment vessel for air cooling to prevent hydrogen explosions, as was done in Chernobyl, instead of injecting water in the melted core." This proposal contains a difficult problem in its own way, and cannot be agreed easily. Nevertheless, Unit 1 temporarily fell into this condition.

Had water injection with fire engines not been conducted, hydrogen explosions would not have occurred at Unit 1. If so, how would the accident have proceeded?   It is presumable, but beyond the scope of this book. I will leave it to future research.

A heap of debris resulting from the core melt in the Chernobyl accident fused the concrete shield at the bottom of the reactor vessel and dropped on the floor of the first basement. Corium, a mixture of molten core materials and concrete, melted in the heap of debris with the decay heat and flew out of the debris three times. It then fused air-conditioning ducts and dropped to the lower floor twice, before finally flying 50 m along the hallway of the reactor building and congealing. Come to think of it, the decay heat of the molten core spreading on the floor was around 0.5 % in all. Once spread, it would not be possible for the molten core to melt the general reinforced concrete floor continuously and bore holes, because the heat diffused from the expanded surface cooled and hardened the molten core.

However, it is good for establishing more definite countermeasures for these situations. As I mentioned in the column in Sect. 2.7 of Chap. 2, Part I, the molten core cannot permeate deep in the reinforced concrete floor because of the high heat conduction of the reinforcing steel. This suggests that, when the reactor building is constructed, the reinforcing steel used in the concrete floor of the containment vessel (DW portion) should be designed to radiate heat by exposing it to ambient air, or welding it onto the wall of the SC before being installed. Subsequently, the heat of the molten core which drops to the floor of the containment vessel will flow toward the outside atmosphere or the water in the SC, and you need not worry about the heat-originated holes at the bottom of the containment vessel.

Research into cooling by radiation heat must be planned in relation to SC water combined with reinforcing steel in the floor of the containment vessel. I am looking forward to future progress.

9.4   Toward More Robust Safety Systems

The National Nuclear Regulatory Authority strengthens regulations by increasing the level of initiators of the natural phenomenon, and request additional facilities of the plant site. However, that may not be adequate to assure nuclear safety in real sense. As discussed in Chap. 6, Sect. 6.2, countermeasures to the threats that may cause common cause failure should be discussed in the international collaboration and be applied to the reactor facilities. Concerning the threats that may exceed them, such as natural disasters beyond our scope, unprepared terrorism attack or whatever unexpected should be explored based on the "mitigation safety systems against disaster (MISSAD)" considerations as discussed in detail in Chap. 6, Sect. 6.3.

The overall nuclear safety can be ensured by more comprehensive concept than safety designs, and this is why I propose MISSAD concept. How to mitigate the disaster is not limited to the countermeasures conducted in the territory of plant site, nor is limited to the facility or components.

As a simile, a traffic safety is supposed. Just as overall traffic safety depends not only on the safety of the car itself, but it also depends on the social infrastructures such as roads, traffic signals, service stations located at appropriate intervals, etc., where the car is the power plant and the infrastructures are the MISSAD.

Consideration for natural disasters is undeniably insufficient in power plants in Japan, but after the Fukushima accident, nuclear power stations nationwide have promoted reinforced measures including the employment of power source cars, watertight distribution panels, and the installation of additional temporary distribution panels.

Among these, distributed arrangement of the power source cars outside the plant site is an example of measures in a category of MISSAD. It may be also very useful as effective measures against the antiterrorism warned by B5b in the U.S.

A high-speed power source ship may be viable as another mean of MISSAD in a country like Japan where many nuclear power plants are located near the sea. The power source vessel may be used as a command center or a haven for evacuees in addition to the power source.

The MISSAD are intended to reduce damage caused by disasters and utilize variety of means and methods in countless combinations. They are initiated under human judgement and actions and are totally opposite to automatically operating common safety measures.

Unlike worldwide NPP safety design guideline, the MISSAD varies depending on the site environment of each NPP. The evacuation plan for each NPP is a kind of MISSAD.

Some are achieved via global collaboration, some by nations, and some by municipalities and power stations according to siting conditions. Proper classification and categorizations, and written policies specifying the roles and responsibilities of the parties concerned may be themes of future study of international organization.

It is obvious that we have to think of MISSAD, but what is required more specifically as organization, as communication network, as facilities or as emergency stocks are subject to discussions hereafter.

Reference

1. Analysis of the TEPCO Fukushima Daiichi NPS Accident (Interim), October 2014. https://www.nsr.go.jp/data/000067237.pdf

Chapter 10
At the End of This Study

These notes conclude my study. Whenever a new fact becomes clear, it should be studied and analyzed one by one carefully, and I hope that the parties engaged in nuclear power generation will compile their study results that supersede this book. I also believe that sharing of information among the worldwide experts is necessary for the future improvement of the plant safety. From this view point, transparent and accurate information dissemination is essential.

10.1 Careful Studies Required to Clarify What Actually Happened

As noted in the introduction, what motivated me to write this book was my attendance at the meeting with the U.S. Science Academy's team for investigating the Fukushima accident. I attended the meeting because I thought it was intended to investigate the facts of the accident but most of its contents, aside from the questions and answers to confirm the facts, comprises an exchange of opinions regarding how computers were used to clarify the accident and similar topics. The meeting did not discuss fuel behavior (reactor core behavior) during the accident, which had been the main theme in the joint safety research studies conducted by the U.S., Germany and Japan.

I felt it might be impossible to clarify what actually occurred in the accident during the period from the core melt to explosions under circumstances like those at this meeting. Accordingly, based on my past studies on fuel behavior (core behavior) during the accident, in which I myself engaged in my youth as well as on the records of the TMI and Chernobyl accidents, I checked data from the Fukushima accident carefully against them and obtained these result as provided in this book.

The following are why I consider computational codes cannot provide adequate solutions.

© Springer Japan 2015

M. Ishikawa, *A Study of the Fukushima Daiichi Nuclear Accident Process*,

DOI 10.1007/978-4-431-55543-8_10

The development of computers has triggered the current trend whereby most design calculations are performed centering on computational codes. Miscalculations have decreased and calculation results generally match experiment results, which has facilitated the task of designing. This is very desirable progress but certain points should still be borne in mind.

Reliable calculation results can only be obtained when the contents of computational code and the target phenomenon agree. I often encounter researchers distressed over discrepancies between a phenomenon and analytical results, which is like putting the cart before the horse. Using computational codes is often problematic and most related problems arise from not completely grasping the contents of computational codes before using them.

It should be noted that the complex phenomena in the Fukushima accident, including its occurrence and progress, remain unclear. To clarify them, opening the reactor heads, which is estimated to take about a decade, and observing the inside of the reactors will be necessary. Without such observation, a final confirmation will be impossible. Accordingly, a computational code applicable to the analysis of the Fukushima accident does not yet exist. Even if findings collected from the TMI accident and previous events having occurred were modeled by extrapolation, such models would not be applicable to the Fukushima accident, which involved peculiar new events exceeding such accident and events in intensity and thus unable to be handled with current safety analysis codes.

As a trend, however, many young engineers in these days cannot work without computers, so they jump to their computers even before gaining insights into accident phenomena. They try to fine-tune the computer inputs radically to make the computational results match the measured accident data forcibly, thereby satisfying themselves with an artificial consistency. Such distorted analysis compromises correct fact-finding. Computational results and measurement data may partially match but coherence to the entire phenomenon is lost, resulting in a deadlock.

What we can do now to determine every fact of the Fukushima accident are to precisely diagnose phenomena based on various facts and data to clarify how the accident occurred as well as implement verification and deliberation while ensuring consistency with all events in the accident. Based on this, the current safety analysis codes should be improved. I hope young researchers will devote utmost efforts to such work and also draw on my study results.

What is the actual fuel behavior dominating the core melt situation during a nuclear accident? This cannot be described in a single phrase. Because there are various accident situations, fuel behavior varies accordingly and each accident has its own specific characteristics. Moreover, fuel behavior varies depending on the position of the fuel rod in the reactor core. This was indicated in the TMI accident, where most of the fuel rods melted but the core-peripheral fuel rods generally retained their original shape.

Accordingly, clarifying the actual fuel behavior seems almost desperate. Nuclear reactor accidents can be broadly divided into the following three types: loss of coolant accidents (LOCAs) caused by cooling water loss, reactivity-initiated accidents (RIAs), where reactor power rapidly increases, and power cooling mismatches

(PCMs) as mentioned previously. Much of the fuel behavior during these three typical accidents has been clarified through experiments performed by the U.S., Germany, Japan, and France. By basically grasping these three patterns of fuel behavior, core melt behavior during an accident can be clarified [1].

10.2   The Government's Failure to Support

On July 9, 2013, when I was writing this book, the Director of the Fukushima Daiichi NPS Masao Yoshida, who had led the handling of the accident, passed away. As commander of the site, he endeavored to bring the accident under control. I pray his soul may rest in peace.

During these studies, I sympathized with Director Yoshida, while considering his feelings during the hardship at the Fukushima Daiichi NPS. As he himself admitted, he may have made some mistakes and reviewing the accident, I find some measures that should have been taken. However, given circumstances where neither electricity nor water was available, handling the accident perfectly must have been impossible.

Without water and electricity, Director Yoshida decided to inject seawater into the reactors from a fire engine. I think this decision was not the procedure he had had in mind from the beginning of the accident. Once seawater is injected into a reactor, the nuclear power plant can never be reused. He must have felt as if he were killing himself and his children. It was a heartbreaking last resort for the director, who had watched the explosion at Unit 1. All the NPP staff members moved intently to implement his, but must have been crying deep inside.

Accordingly, Director Yoshida had no other alternative, but the Government could have taken some measures. When electricity is unavailable due to a blackout, he/she must do anything to get it back.

In the case of Unit 3, by the time the reactor was still very cold, more than one and a half days had passed. If electricity had been supplied from outside in one way or another by this time, Unit 3 could have been brought to a cold shutdown.

Given the capability and technologies of Japan, electricity supply was possible. For example, a vessel of the Ministry of Defense could have been dispatched to the accident site to supply electricity. The tsunami had subsided by then and the NPS has a quay for unloading cargo. Certainly, it is not easy to supply electricity from a vessel but the situation was an emergency where a core melt may have been imminent. The garbage generated after the tsunami floating on the sea is no more than an excuse. If only the Government had been more willing, it could have taken any measure.

The fact the Fukushima Daiichi NPS would have been saved had electricity been available is proven by the fact the Onagawa NPS and the Fukushima Daini NPS safely achieved cold shutdown. In addition, one of the emergency diesel generators installed at the Fukushima Daiichi site, fortunately survived the accident. The operators used this diesel generator to supply power to Units 5 and 6 and brought their reactors to a cold shutdown state.

Had the former Prime Minister Naoto Kan calmly devoted himself to logistically supporting the NPS without persisting on site visits by plane, the Units 2 and 3 could have been saved, which is my conclusion. What on earth were the Government leaders gathering in the Prime Minister's official residence, TEPCO's leaders, the Nuclear Safety Commission's members and government officials doing back then? Frankly speaking, they were utterly useless. I do not know how best to vent my fury toward them.

The accident took place because of the natural disasters but spread as a man-made disaster. We should bear this in mind and devise future safety measures.

10.3   More Information Required from TEPCO for More Accurate Analyses

One thing I have noticed through these studies is that information from related parties has been quite scarce compared to that from the TMI and Chernobyl accidents.

I resolved to start studying and writing this book in around February 2013. Because I could estimate the outline of the accident progress from the beginning, I took the work lightly at first. However, once I started working on data published by TEPCO, it proved quite hard. For example, with regard to the core melt due to the reaction between fuel cladding (zirconium) and water, this ended about 2 min after water injection in the TMI accident, while there was a considerable time lag between water injection and core melt at each unit in the Fukushima Daiichi NPS. In the case of the latter, the water-zirconium reaction occurred long after the water injection and triggered the reactor building explosions. What caused this difference? To answer such questions one by one consistently in this book, I repeatedly read TEPCO's report, from cover to cover, but information helping me solve the above question remained very scarce.

In mid-April 2013, when I was contemplating this question, Mr. Akiyoshi Minematsu, former TEPCO Vice President, told me in a meeting that the water had been injected from a fire engine into a pipe connected to the inner wall of the reactor pressure vessel, which answered most of my question. I had thought the injected water was sprayed over the top of the reactor core. In fact, the injected water entered from the bottom of the reactor core slowly, which created a core melt situation unlike what I estimated previously. His explanation blew away the cobwebs in my head. Just a piece of specific information often plays a critical role in an accident investigation.

When I worked to clarify the details of the Chernobyl accident long ago, records and statements by related parties were very helpful. For example, a remark "sparks were dancing above the plant like devils" indicated that cooling water was spouting from the reactor, and a remark "dancing shield plug" was a convincing testimony indicating a time lag of a few minutes between the reactor power excursion and the explosions (see one of my works *Reactor Power Excursion*) [2].

What bothered me during my studies was the lack of response from TEPCO concerning accident situations like these and actions that had been taken during the accident. Therefore I could not determine specific facts that could be used as supporting evidence. This is probably inevitable because there may be various circumstances involving the Government and given the Japanese culture and customs but excessive information control results in lost opportunities to improve nuclear safety.

On December 13, 2013, when I was finalizing this book, TEPCO announced a new report on the situation of seawater injected in Units 1 to 3. This information was really useful. The report said that the pipe that conveyed seawater from the fire hose to the reactor had some branches, from which the seawater flowed out, meaning the injected seawater did not reach the reactor as expected. In other words, because of the high containment vessel pressure, seawater did not enter the reactor adequately and flowed out of the branched pipes. In the explanation regarding the "oozy" burning reaction that lasted for 20 h at Unit 3, I stated that "I ignored the calculation regarding water injection because this phenomenon is unexplainable without thinking this way" but injected water flowing out of the pipe branches changes the story completely. As I estimated, the amount of water sent into the reactor was small, which is reflected in the fact that the "oozy" burning reaction lasted for an extended period. This new information supported my studies and deepened my confidence in my study results.

This example shows the importance of transmitting various accident-related information honestly and actively.

10.4   The Need to Provide Precise Information to the World

In spring 2012, when I started writing this book, a vehement antinuclear power movement also emerged in Taiwan. From what I heard, it was triggered by agitation by a Japanese activist who said "Should a nuclear accident like that in Fukushima occur in Taiwan, 30 thousand people would die immediately and 7 million people would die of cancer." Pushed by the heated-up antinuclear movement, president Ma Ying-Jeou decided to limit the operable period of NPPs to 40 years, taking an example from Japan, and decided to hold a referendum to decide whether the construction of Unit 4 should be continued.

Jinshan Nuclear Power Plant Unit 1, which is the first unit built in Taiwan, has been operating for about 35 years, which means it is operable for another 5 years before reaching the 40-year operation limit. A Taiwanese engaged in nuclear energy field visited me for advice saying a decommissioning plan needs to be submitted to the Taiwan Government 3 years before the operation limit. He asked me to hold study meetings to share Japanese experiences in decommissioning planning and work and give a lecture related to the circumstances of the Fukushima accident. As this suggests, information on the Fukushima accident owned by Japan has barely reached the world.

I have visited Taiwan in June, 2013 and gave a lecture there. At the entrance of the lecture hall, many people purporting to be victims of the Fukushima accident were holding an antinuclear gathering. I heard that they were doing so almost daily. Many of the questions asked by lecture attendees were how the Japanese activists (with their name specifically mentioned in the questions) participating in the anti-nuclear gatherings in Taiwan were performing activities in the atomic energy society in Japan as well as the authenticity of the contents of his speech.

In particular, they did not know the fact that, despite the 3-unit core melt in the Fukushima Daiichi NPS, no-one died of the nuclear disaster. Even president Ma Ying-Jeou, whom I met the next day, asked me "Is it true that no-one died in the nuclear disaster?" to confirm the authenticity of my previous statement.

When I said in the lecture that Unit 3, where the hydrogen explosions occurred, was loaded with mixed uranium and plutonium oxide fuel (MOX fuel), a stir rippled through the hall because they heard it for the first time. After hearing my speech diametrically opposing what the antinuclear activists had said, the attendees seemed outraged with them. After the lecture, some attendees even asked me questions to check the authenticity of my lecture contents.

About a decade ago, a vigorous antinuclear movement emerged over the use of MOX fuel in Japan. The activists insisted that a "MOX fuel core melt would kill hundreds of thousands of people through plutonium inhalation." Looking back on the Fukushima Daiichi NPS accident, however, such insistence proved completely wrong. Even though the reactor cores melted and hydrogen explosions occurred at three units, no-one died.

Immediately after I returned to Japan from Taiwan, the comment "the number of deaths in the Fukushima accident was zero" by a politician was criticized. The reason was that this comment disregarded the people who died during their refugee lives, but is this comment really problematic?

Such a criticism is an attempt to cover up the facts of the accident and does nothing to help improve nuclear safety. This also reveals an unwillingness to face the facts of the accident squarely, which is hindering the precise transmission of information and propagating distrust in Japan.

10.5   Fukushima Reconstruction with Cool-Headed Judgments

These days, the problem of contaminated water has become a major topic.

Although contaminated water has been significantly cleaned up, it has been stored in tanks because of local people's disagreement to its release into the sea. Accordingly, even the vast Fukushima Daiichi NPS site has filled up with storage tanks.

Decontaminating the contaminated groundwater and storing it in tanks is wasteful and meaningless. Decontaminating the groundwater to the radioactive concentration level allowed by international standards and then releasing it to the ocean would solve the problem of contaminated water. The Government should strive to solve this problem.

I also hope the local people consider this problem calmly. Water with a radioactive concentration of less than the national radioactive release standard is equivalent in purity to water released into the sea under normal operation. Releasing the decontaminated water into the sea is likely to cause fewer problems in future and expenses used to additionally build tanks could be diverted to local reconstruction.

Refugee lives and compensation problems are compounding problems for the people in Fukushima. Apart from the responsibility of the accident, I hope they further consider the bravery of TEPCO employees who ventured to squarely face the nuclear accident in the Fukushima Daiichi site. The employees challenged the nuclear disaster while hoping to avoid inconvenience to the residents in the vicinity. Even now, they are striving under severe conditions to solve ongoing problems.

Three years have already elapsed since the accident and the radiation level in many areas, excluding those significantly affected in the vicinity of the NPS, has declined to a level having no effect on human health. As I stated in Part II, Chap. 4, a decontamination level of 1 mSv is too strict to be accepted even by international standards. From the perspective of international standards, which have developed based on scientific judgment, an environment under which evacuees can return home is already in place.

Although the choice of return home or no depends on the individual judgment of each evacuee, the hope for the return home and restart of job in the hometown remains open. It is natural for the local residents to hold a grudge against TEPCO, which caused the accident, but hatred does not create hopes. I really hope the local residents will calmly observe the current situation, consider their countrymen striving to challenge the radiation in the Fukushima Daiichi site, and contribute to future reconstruction while supporting one another.

References

1. Nuclear Safety Research Association (2013) The behaviour of light water reactor fuel [Japanese version only]
2. Ishikawa M (1996) Reactor power excursion. Nikkan Kogyo Shimbun, Ltd. [Japanese version only]

Postscript

I should have published this book much earlier. As I noted in the introduction, I initially thought clarifying the Fukushima accident would be the task of younger researchers and did not dare to engage myself in it. It was the end of November 2012 when I changed my mind; concerned about the poor state of the accident investigation. When it comes to writing a book, however, specific knowledge regarding the Fukushima Daiichi NPS on the one hand and clear knowledge regarding the TMI accident on the other are necessary.

Fortunately, Mr. Nobuyuki Kitamura, once a member of the former Japan Nuclear Technology Institute (JANTI [current Japan Nuclear Safety Institute]) since its establishment, had experience of working in the Fukushima Daiichi and Daini NPSs for about 15 years, so I was able to exchange opinions with him from the beginning of the Fukushima accident. Mr. Kitamura provided me with all the knowledge regarding the Fukushima Daiichi NPS necessary to precisely understand the phenomena there. Moreover, I asked him to check the draft of this book to see whether it was correct compared to the actual site situation. Although the title includes the word "study" because no data was recorded during the accident due to blackout, I am confident that all I have written in this book is the truth. Without advice and checking by Mr. Kitamura, I could not have written this book so confidently.

With regard to the TMI accident, I was very fortunately able to utilize the experiences of Mr. Nobuyuki Onishi, who had temporarily transferred from the Shikoku Electric Power Co, Inc. to the JANTI just before the Fukushima accident and had been engaged in a TMI accident investigation for 2 years in the U.S. as the Japan & U.S. WR Research Committee's TMI delegate. He retains a mental record of detailed and precise TMI-related data. My rough knowledge regarding the TMI accident was precisely reinforced and corrected by him and I have repeatedly reaffirmed that knowledge of events in the TMI accident and their precise interpretation are the bases for verifying the core melts and explosions at the Fukushima Daiichi

© Springer Japan 2015
M. Ishikawa, *A Study of the Fukushima Daiichi Nuclear Accident Process*,
DOI 10.1007/978-4-431-55543-8

NPS as the central themes of this book. Moreover, Mr. Onishi edited and proofread the draft of this book unceasingly on my behalf, given my unfamiliarity in handling electronic devices.

Without the cooperation of these two persons, it would have been impossible for me to write this book. I would like to express my deep gratitude to both of them.

I was able to grasp the details of the accident almost entirely by around the end of April 2013, whereupon I started drafting this book. Before completing the draft, I showed it to the two persons to obtain their comments and then rewrote it, which I repeated until the end of August. Interestingly, writing itself exposes misunderstandings and inconsistencies. Writing about an accident involving one reactor would not be so bothersome, but when it comes to writing about one involving as many as three reactors, it is a different story because a single misunderstanding or inconsistency affects the entire consistency. I repeatedly went back to square one to review all the reactors from scratch, which continued until the end of the summer. By the end of September, I managed to complete the draft.

However, I still remained in the dark regarding one question. It was the time of the initial background radiation level spike I mentioned in Part II, Chap. 4 and I was at my wit's end over the fact that, although the Unit 1 vent valve had been opened at 10 a.m. the day after the start of the accident, the radiation level had started to increase earlier at 4 a.m., which puzzled me. As many people suggested, if the radioactive substances had leaked from a containment vessel electrical penetration or an equipment hatch, it would have naturally entailed hydrogen leakage and therefore it is unnatural that the hydrogen explosion at Unit 1 occurred at only the 5th floor. For example, at Unit 3, where hydrogen leaked to the lower floors, explosions also occurred on those floors. Eventually, it emerged that I had overlooked the explanation regarding this question in the report. It was not until I found the cause of this problem (see Part II, Chap. 4, Sect. 4.1.) that I had confidence about the hydrogen gas leakage routes and resolved to publish this book.

At the end of October 2013, when the draft and drawings were mostly completed, I asked the following persons to examine and proofread my draft: Dr. Toshio Fujishiro, with whom I had conducted fuel behavior experiments using a nuclear safety research reactor (NSRR) during my stint in the Japan Atomic Energy Research Institute (JAERI), Mr. Masayuki Akimoto, who developed safety analysis codes during my stint in the Division of Systems Safety and Analysis, and Mr. Kenichi Sugiyama, a former professor at the Hokkaido University and my successor. Subsequently, I asked the following persons to check my draft: persons engaged in the handling of the accident at the Fukushima Daiichi NPS and an old friend of mine Mr. Harukuni Tanaka, who formerly worked as the chief engineer of reactors at Units 1 and 2 of the Fukushima Daiichi NPS. Accordingly, I am confident that the outlines of the accident provided in this book are generally correct apart from the fine details.

In addition, a new report was published by TEPCO regarding the progress of seawater injection on December 13, 2013 when I was finalizing the draft. The report solved what I had considered an inconsistency between the Unit 3 reactor water level data and the amount of seawater injection, about which I had been most uncertain. I was finally able to have certainty about the "oozy" reaction I had estimated.

When I was finalizing the draft, I had the opportunity to watch video footage explaining the Fukushima's core melts, which was made public on the website of the Institute for Radioprotection and Nuclear Safety (IRSN). The footage described the core melting heat as the result of water-zirconium reaction. This is exactly what occurred. I was impressed by the research capability of France and simultaneously regretted the research level in Japan.

With regard to the core melt situation, however, the video footage assumed a scenario where the liquefied molten core trickled down and melted the reactor pressure vessel bottom, which was the same as the scenario aired by NHK. As I noted in Part II, Chap. 5, this scenario seems wrong in terms of the quantity of heat but I want to examine this with researchers after related reports are issued in the near future. I hope the release of this book creates an opportunity to clarify and solve the questions of the Fukushima accident.

To make this book readable, I asked Mr. Hirotsugu Hayashi of the Japan Nuclear Safety Institute (JANSI) to review it and add annotations and make corrections to improve the clarity of the figures and tables together with related sentences. In the selection of the publisher of this book, Mr. Yoshihiko Kawai, one of my close associates since long time ago, made various arrangements. Moreover, Ms. Noriko Jindo and Mr. Daisuke Saisho (Denki Shimbun) made significant efforts to edit this book.

Accordingly, this book was completed with cooperation from these prominent figures engaged in the nuclear energy field, to whom I express deep gratitude for their significant efforts.

March 2014   Michio Ishikawa

One year after my plan about the translation of this book, the English edition is now published from Springer.

First, I would like to thank to the advocates who pulled the trigger of this English edition project, Mr. Akito Arima, Mr. Kumao Kaneko, Mr. Tetsuo Takeuchi, Mr. Masao Takuma, Mr. Masao Nakamura, and Mr. Teruaki Masumoto. I would also like to express my hearty thanks to Japanese BWR vendors and some other companies that supported this project.

Finally, I express my special thanks to Dr. Shinzabro Matsuda, prominent researcher of the Tokyo Institute of Technology and my skiing companion, who helped me in making abstracts and proofreading of the text with full heart as if he were the author, in cooperating with Mr. Nobuyuki Onishi, whom I already mentioned. I also thank to Mr. Minoru Mochizuki and D&Y Co.,Ltd. for translation of Part I and II, respectively.

July 2015   Michio Ishikawa

About the Author

Michio Ishikawa was born in Takamatsu City, Kagawa prefecture in 1934. He graduated from the Department of Mechanical Engineering, The University of Tokyo with a bachelor of engineering and joined the Japan Atomic Energy Research Institute (JAERI) in 1957. He was engaged in constructing and operating the Japan Power Demonstration Reactor (JPDR), which succeeded in generating electricity through nuclear energy for the first time in Japan in 1963. After joining a special power excursion reactor test (SPERT) at the National Reactor Test Station (NRTS) in Idaho, he developed and performed an NSRR experiment plan concerning reactivity-initiated accidents. Starting from 1985, he led the first decommissioning of the nuclear power reactor in Japan. After working as vice president of the JAERI's Tokai Research Establishment, he became a professor of the Engineering Department of Hokkaido University. After retirement, he served as technical counselor for the Japan Nuclear Energy Safety Organization (JNES). In April 2005, he became president of the former Japan Nuclear Technology Institute (JANTI, which is the current Japan Nuclear Safety Institute [JANSI]), and in April 2008 to September 2012, served as the top advisor to the JANTI.

During 1973–2004, he held positions such as the nuclear safety advisor of the former Science and Technology Agency (current Ministry of Education, Culture, Sports, Science and Technology), the nuclear power generation technology advisor of the Ministry of Economy, Trade and Industry (METI), the expert advisor for the Central Disaster Prevention Council, and the Japanese representative of the committees of the International Atomic Energy Agency (IAEA) and the Organization for Economic Cooperation and Development (OECD/NEA).

Main works: *Reactor Dismantling* (Kodansha Ltd. 1993) [Japanese version only]

Reactor Power Excursion (Nikkan Kogyo Shimbun, Ltd. 1996) [Japanese version only]

Views on Nuclear Energy (The Denki Shimbun, Japan Electric Association Newspaper Division 2005) [Japanese version only]

© Springer Japan 2015   231
M. Ishikawa, *A Study of the Fukushima Daiichi Nuclear Accident Process*,
DOI 10.1007/978-4-431-55543-8